A PLANET OF VIRUSES

CARL ZIMMER

A Planet of
VIRUSES

The University of Chicago Press | Chicago and London

CARL ZIMMER writes about science for the *New York Times* and other publications and is the author of eight books, including *Parasite Rex*, *Soul Made Flesh*, and *Microcosm*. He is a lecturer at Yale University, where he teaches writing about science and the environment, and visiting scholar at the Science, Health, and Environment Reporting Program at New York University's Arthur L. Carter Journalism Institute.

The University of Chicago Press, Chicago 60637
The University of Chicago Press, Ltd., London
© 2011 by The Board of Regents of the University of Nebraska
All rights reserved. Published 2011.
Printed in the United States of America

20 19 18 17 16 15 14 13 12 11   1 2 3 4 5

ISBN-13: 978-0-226-98335-6   (cloth)
ISBN-10: 0-226-98335-8   (cloth)

The essays in this book were written for the World of Viruses project, funded by the National Center for Research Resources at the National Institutes of Health through the Science Education Partnership Award (SEPA) Grant No. R25 RR024267 (2007–2012). Its content is solely the responsibility of the authors and does not necessarily represent the official views of NCRR or NIH. Visit http://www.worldofviruses.unl.edu for more information and free educational materials about viruses. World of Viruses is a project of the University of Nebraska–Lincoln.

Library of Congress Cataloging-in-Publication Data

Zimmer, Carl, 1966–
    A planet of viruses / Carl Zimmer.
        p. cm.
    Includes bibliographical references and index.
    ISBN-13: 978-0-226-98335-6 (cloth : alk. paper)
    ISBN-10: 0-226-98335-8 (cloth : alk. paper) 1. Viruses. I. Title.
    QR360.Z65 2011
    362.196'9—dc22

                                    2010036742

♾ This paper meets the requirements of ANSI/NISO Z39.48-1992 (Permanence of Paper).

To Grace, my favorite host

Contents

Foreword

Viruses wreak chaos on human welfare, affecting the lives of almost a billion people. They have also played major roles in the remarkable biological advances of the past century. The smallpox virus was humanity's greatest killer, and yet it is now one of the only diseases to have been eradicated from the globe. New viruses, such as HIV, continue to pose new threats and challenges.

Viruses are unseen but dynamic players in the ecology of Earth. They move DNA between species, provide new genetic material for evolution, and regulate vast populations of organisms. Every species, from tiny microbes to large mammals, is influenced by the actions of viruses. Viruses extend their impact beyond species to affect climate, soil, the oceans, and fresh water. When you consider how every animal, plant, and microbe has been shaped through the course of evolution, one has to consider the influential role played by the tiny and powerful viruses that share this planet.

Carl Zimmer wrote these essays for the World of Viruses project as part of a Science Education Partnership Award (SEPA) from the National Center for Research Resources (NCRR) at the National Institutes of Health (NIH). World of Viruses was created to help people understand more about viruses and virology research through radio documentaries, graphic stories, teacher professional development, mobile phone and iPad applications, and other materials. For more information about World of Viruses, visit http://www.worldofviruses.unl.edu.

JUDY DIAMOND, PhD
*Professor and Curator, University of Nebraska
    State Museum*
Director of the World of Viruses Project

CHARLES WOOD, PhD
*Lewis L. Lehr University Professor of Biological
    Sciences and Biochemistry*
Director of the Nebraska Center for Virology

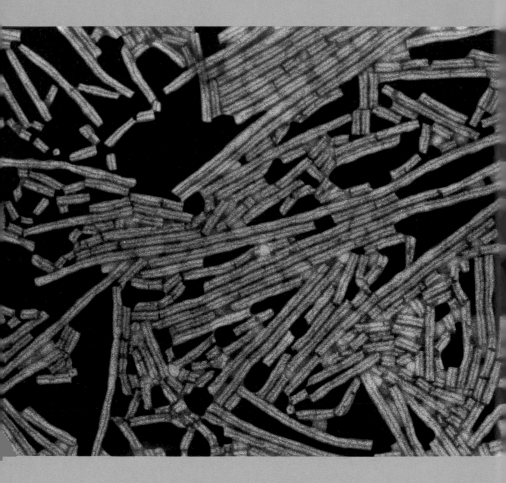

Tobacco mosaic viruses, which cause plant diseases worldwide

Introduction

"A Contagious Living Fluid"
Tobacco Mosaic Virus

Fifty miles southeast of the Mexican city of Chihuahua is a dry, bare mountain range called Sierra de Naica. In 2000, miners worked their way down through a network of caves below the mountains. When they got a thousand feet underground, they found themselves in a place that seemed to belong to another world. They were standing in a chamber measuring thirty feet wide and ninety feet long. The ceiling, walls, and floor were lined with smooth-faced, translucent crystals of gypsum. Many caves contain crystals, but not like the ones in Sierra de Naica. They measured up to thirty-six feet long apiece and weighed as much as fifty-five tons. These were not crystals to hang from a necklace. These were crystals to climb like hills.

Since its discovery, a few scientists have been granted permission to visit this extraordinary chamber, known now as the Cave of Crystals. Juan Manuel García-Ruiz, a geologist at the University of Granada, made the journey and figured out that the crystals formed when volcanoes began to form the mountains 26 million years ago. Subterranean chambers took shape and filled with hot mineral-laced water. The heat of the volcanic magma kept the water at around 136 degrees, the ideal temperature for the minerals to settle out of the water and form crystals. Somehow the water stayed at that perfect temperature for hundreds of thousands of years, allowing the crystals to grow to surreal sizes.

In 2009, another scientist, Curtis Suttle, paid a visit to the Cave of Crystals. Suttle and his colleagues scooped up water from the chamber's pools and brought it back to their laboratory at the University of British Columbia to analyze. When you consider Suttle's line of work, his journey might seem like a fool's errand. Suttle has no professional interest in crystals, or minerals, or any rocks at all for that matter. He studies viruses.

There are no people in the Cave of Crystals for the viruses to infect. There are not even any fish. The cave has been effectively cut off from the biology of the outside world for millions of years. Yet Suttle's trip was well worth the effort. After he prepared his samples of crystal water, he put them under a microscope and saw protein shells loaded with genes. Each drop of cave water may hold two hundred million viruses.

Just about wherever scientists look—deep within the earth, on grains of sand blown off of the Sahara Desert, under mile-thick layers of Antarctic ice—they find viruses. And when they look in familiar places, they find new ones. In 2009, Dana Willner, a biologist at San Diego State University, led a virus-hunting expedition into the human body. The scientists had ten people cough up sputum and spit it into a cup. Five of the people were sick with cystic fibrosis, and five were healthy. Out of that fluid, Willner and her team fished out fragments of DNA, which they compared to databases of the tens of millions of genes already known to science. Before Willner's study, the lungs of healthy people were believed to be sterile. But Willner and her colleagues discovered that all their subjects, sick and healthy alike, carried viral menageries in their chests. On average, each person had 174 species of viruses in the lungs. But only 10 percent of those species bore any close kinship to any virus ever found before. The other 90 percent were as strange as anything lurking in the Cave of Crystals.

The science of virology is still in its early, wild days. Scientists are discovering viruses faster than they can make sense of them. And yet this is a late-blooming youth, for we have known about viruses for thousands of years. We have known them from their effects, in our sicknesses and our deaths. But for centuries we did not know how to join those effects to their cause. The very word *virus* began as a contradiction. We inherited the word from the Roman Empire, where it meant, at once, the venom of a snake or the semen of a man. Creation and destruction in one word.

Over the centuries, *virus* took on another meaning: it signified any contagious substance that could spread disease. It might be a fluid, like the discharge from a sore. It might a substance that traveled mysteriously through the air. It might even impregnate a piece of paper, spreading disease with the touch of a finger. *Virus* only began to take on its modern meaning as the nineteenth century came to a close, thanks to an agricultural catastrophe. In the Netherlands, tobacco farms were swept by a disease that left plants stunted, their leaves a mosaic of dead and live patches of tissue. Entire farms had to be abandoned.

In 1879, Dutch farmers came to Adolph Mayer, a young agricultural chemist, to beg for help. Mayer carefully studied the scourge, which he dubbed tobacco mosaic disease. He investigated the environment in which the plants grew—the soil, the temperature, the sunlight. He could find nothing to distinguish the healthy plants from the sick ones. Perhaps, he thought, the plants were suffering from an invisible infection. Plant scientists had already demonstrated that fungi could infect potatoes and other plants, so Mayer looked for fungus on the tobacco plants. He found none. He looked for parasitic worms that might be infesting the leaves. Nothing.

Finally Mayer extracted the sap from sick plants and injected drops into healthy tobacco. The healthy plants, Mayer discovered, turned sick as well. Some microscopic pathogen must be multiplying inside the plants. Mayer took sap from sick plants and incubated it in his laboratory. Colonies of bacteria began to grow and became large enough that Mayer could see them with his naked eye. Mayer applied the bacteria to healthy plants to see if it would trigger tobacco mosaic disease. It failed. And with that failure, Mayer's research ground to a halt.

A few years later, another Dutch scientist named Martinus Beijerinck picked up where Mayer left off. He wondered if something other than bacteria was responsible for tobacco mosaic disease, something far smaller. He ground up diseased plants and passed the fluid through a fine filter that blocked both plant cells and bacteria. When he injected the clear fluid into healthy plants, they became sick.

Beijerinck filtered the juice from the newly infected plants and found that he could infect still more tobacco. Something in the sap of the infected plants—something smaller than bacteria—could replicate itself and could spread disease. Beijerinck called it a "contagious living fluid."

Whatever that contagious living fluid carried was different from any other kind of life biologists knew about. It was not only inconceivably small but also remarkably tough. Beijerinck could add alcohol to the filtered fluid, and it would remain infective. Heating the fluid to near boiling did it no harm. Beijerinck soaked filter paper in the infectious sap and let it dry. Three months later, he could dip the paper in water and use the solution to sicken new plants.

Beijerinck used the word *virus* to describe the mysterious agent in his contagious living fluid. It was the first time anyone used the word the way we do today. But in a sense, Beijerinck simply used it to define viruses by what they were *not*. They were not animals, plants, fungi, or bacteria. What exactly they were, Beijerinck could not say. He had reached the limits of what nineteenth-century science could reveal.

A deeper understanding of viruses would have to wait for better tools and better ideas. Electron microscopes allowed scientists to see viruses for what they are: particles of a nearly inconceivably small size. For comparison, tap out a single grain of salt from a shaker. You could line up about ten skin cells along one side of it. You could line up about a hundred bacteria. Compared to viruses, however, bacteria are giants. You could line up a thousand viruses alongside that same grain of salt.

Despite the small size of viruses, scientists discovered ways to dissect them and peer inside. A human cell is stuffed with millions of different molecules that it uses to sense its surroundings, crawl hither and yon, take in food, grow, and decide whether to divide in two or kill itself for the good of its fellow cells. Virologists found that many of the viruses they studied were just protein shells holding a few genes. They discovered that viruses can replicate themselves, despite their paltry genetic instructions, by hijacking other forms of life. They could see viruses inject their genes and proteins into a host cell, which they manipulated into producing new copies of the virus. One virus might go into a cell, and within a day a thousand viruses came out.

Virologists had grasped these fundamental facts by the 1950s. But virology did not come to a halt. For one thing, virologists knew little about the many different ways in which viruses make us sick. They didn't know why papillomaviruses can cause horns to grow on rabbits and cause hundreds of thousands of cases of cervical cancer each year. They didn't know what made some viruses deadly and others relatively harmless. They had yet to learn how viruses evade the defenses of their hosts and how they evolve faster than anything else on the planet. In the 1950s they did not know that a virus that would later be named HIV had already spread from chimpanzees into our own species, or that thirty years later it would become one of the greatest killers in history. They could not have dreamed of the vast numbers of viruses that exist on Earth; they could not have guessed that most of the genetic diversity of life can be found in virus genes. They did not know that viruses help produce much of the oxygen we breathe and help control the planet's thermostat. And they certainly would not have guessed that the human genome is partly composed from thousands of viruses that infected our distant ancestors, or that life as we know it may have gotten its start four billion years ago from viruses.

Now scientists know these things—or, to be more precise, they know *of* these things. They now recognize that from the Cave of Crystals to the inner world of the human body, this is a planet of viruses. Their understanding is still rough, but it is a start. So let us start as well.

OLD COMPANIONS

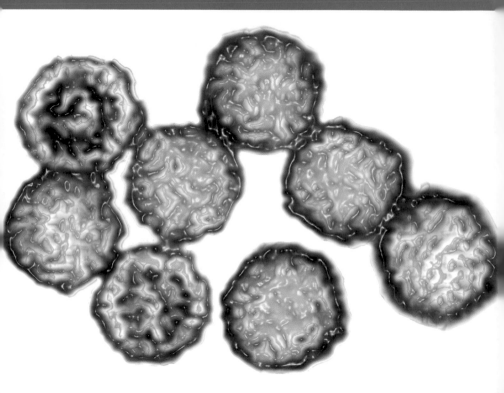

Rhinoviruses, the most common cause of colds

The Uncommon Cold

Rhinovirus

Around 3,500 years ago, an Egyptian scholar sat down and wrote the oldest known medical text. Among the diseases he described in the so-called Ebers Papyrus was something called *resh*. Even with that strange sounding name, its symptoms—a cough and a flowing of mucus from the nose—are immediately familiar to us all. *Resh* is the common cold.

Some viruses are new to humanity. Other viruses are obscure and exotic. But human rhinoviruses—the chief cause of the common cold, as well as asthma attacks—are old, cosmopolitan companions. It's been estimated that every human being will spend a year of his or her life lying in bed, sick with colds. The human rhinovirus is, in other words, one of the most successful viruses of all.

Hippocrates, the ancient Greek physician, believed that colds were caused by an imbalance of the humors. Two thousand years later, the physiologist Leonard Hill argued in the 1920s that they were caused by walking outside in the morning, from warm to cold air. The first clue to the true cause of colds came when Walter Kruse, a German microbiologist, had a snuffly assistant blow his nose and mix the mucus into a salt solution. Kruse and his assistant purified the fluid through a filter and then put a few drops into the noses of twelve of their colleagues. Four of them came down with colds. Later, Kruse did the same thing to thirty-six students. Fifteen of them got sick. Kruse compared their outcomes to thirty-five people who didn't get the drops. Only one of the drop-free individuals came down with a cold.

Kruse's experiments made it clear that some tiny pathogen was responsible for the cold. At first, many experts believed it was some kind of bacteria, but Alphonse Dochez ruled that out in 1927. He filtered the mucus from people with colds, the same way Beijerinck had filtered tobacco plant sap thirty years before, and discovered that the bacteria-free fluid could make people sick. Only a virus could have slipped through Dochez's filters.

It took another three decades before scientists figured out exactly which viruses had slipped through. Known as human rhinoviruses (*rhino* means nose), they are remarkably simple, with only ten genes apiece. (We have twenty thousand.) And yet that haiku of genetic information is enough to let the human rhinovirus invade our bodies, outwit our immune system, and give us colds.

The human rhinovirus spreads by making noses run. People with colds wipe their noses, get the virus on their hands, and then spread the virus onto door knobs and other surfaces they touch. The virus hitches onto the skin of other people who touch those surfaces and then slips into their body, usually though their nose. Rhinoviruses can invade the cells that line the interior of the nose, throat, or lungs. They trigger the cells to open up a trapdoor through which they slip. Over the next few hours, a rhinovirus will use its host cells to make copies of its genetic material and protein shells to hold them. The host cell then rips apart, and the new virus escapes.

Rhinoviruses infect relatively few cells, causing little real harm. So why can they cause such miserable experiences? We have only ourselves to blame. Infected cells release special signaling molecules, called cytokines, which attract nearby immune cells. Those immune cells then make us feel awful. They create inflammation that triggers a scratchy feeling in the throat and leads to the production of a lot of mucus around the site of the infection. In order to recover from a cold, we have to wait not only for the immune system to wipe out the virus but also to calm itself down.

The Egyptian author of the Ebers papyrus wrote that the cure for *resh* was to dab a mixture of honey, herbs, and incense around the nose. In seventeenth-century England, cures included a blend of gunpowder and eggs and of fried cow dung and suet. Leonard Hill, the physiologist who believed a change of temperature caused colds, recommended that children start their day with a cold shower. Today, doctors don't have much more to offer people who get colds. There is no vaccine. There is no drug that has consistently shown signs of killing the virus. Some studies have suggested that taking zinc can slow the growth of human rhinoviruses, but later studies failed to replicate their results.

In fact, some treatments for the cold may be worse than the disease itself. Parents often give their children cough syrup for colds, despite the fact that studies show it doesn't make people get better faster. But cough syrup also poses a wide variety of rare yet serious side effects, such as convulsions, rapid heart rate, and even death. In 2008, the Food and Drug Administration warned that children under the age of two—the people who get colds the most—should not take cough syrup.

Another popular treatment for the cold is antibiotics, despite the fact that they only work on bacteria and are useless again viruses. In some cases, doctors prescribe antibiotics because they're not sure whether a patient has a cold or a bacterial infection. In other cases, they may be responding to pressure from worried parents to do *something*. But unnecessary prescriptions of antibiotics are a danger to us all, because they foster the evolution of increasingly drug-resistant bacteria in our bodies and in the environment. Failing to treat their patients, doctors are actually raising the risk of other diseases for everyone.

One reason the cold remains incurable may be that we've underestimated the rhinovirus. It exists in many forms, and scientists are only starting to get a true reckoning of its genetic diversity. By the end of the twentieth century, scientists had identified dozens of strains, which belonged to two great lineages, known as HRV-A and HRV-B. In 2006, Ian Lipkin and Thomas Briese of Columbia University were searching for the cause of flu-like symptoms in New Yorkers who did not carry the influenza virus. They discovered that a third of them carried a strain of human rhinovirus that was not closely related to either HRV-A or HRV-B. Lipkin and Briese dubbed it HRV-C, and since then, researchers have found that this third lineage is common around the world. From one region to another, the variations in HRV-C's genes are few, which suggests that the virus wasted no time spreading through our species. In fact, the common ancestor of all HRV-C may be just a few enturies old.

The more strains of rhinoviruses scientists discover, the better they come to understand their evolution. All human rhinoviruses share a core of genes that have changed very little as the viruses have spread around the world. Meanwhile, a few parts of the rhinovirus genome are evolving very quickly. These regions appear to help the virus avoid being killed by our immune systems. When our bodies build antibodies that can stop one strain of human rhinovirus, other strains can still infect us because our antibodies don't fit on their surface proteins. Consistent with this hypothesis is the fact that people are typically infected by several different human rhinovirus strains each year.

The diversity of human rhinoviruses makes them a very difficult target to hit. A drug or a vaccine that attacks one protein on the surface of one strain may prove to be useless against others that have a version of that protein with a different structure. If another strain of human rhinovirus is even a little resistant to such treatments, natural selection can foster the spread of new mutations, leading to much stronger resistance.

Despite the diversity of rhinoviruses, some scientists are optimistic that they can develop a cure for the common cold. The fact that all strains of human rhinoviruses share a common core of genes suggests that the core can't withstand mutations. In other words, viruses with mutations in the core die. If scientists can figure out ways to attack the rhinovirus core, they may be able to stop the disease. One promising target is a stretch of genetic material in rhinoviruses that folds into a loop shaped like a clover leaf. Every rhinovirus scientists have studied carries the same clover-leaf structure, which appears to be essential for speeding up the rate at which a host cell copies rhinovirus genes. If scientists can find a way to disable the clover leaf, they may be able to stop every cold virus on Earth.

But should they? Human rhinoviruses certainly impose a burden on public health, not just by causing colds but by opening the way for more harmful pathogens. But the human rhinovirus itself is relatively mild. Most colds are over in a week, and 40 percent of people who test positive for rhinoviruses suffer no symptoms at all. In fact, human rhinoviruses may offer some benefits to their human hosts. Scientists have gathered a great deal of evidence that children who get sick with relatively harmless viruses and bacteria may be protected from immune disorders when they get older, such as allergies and Crohn disease. Human rhinoviruses may help train our immune systems not to overreact to minor triggers, instead directing their assaults to real threats. Perhaps we should not think of colds as ancient enemies but as wise old tutors.

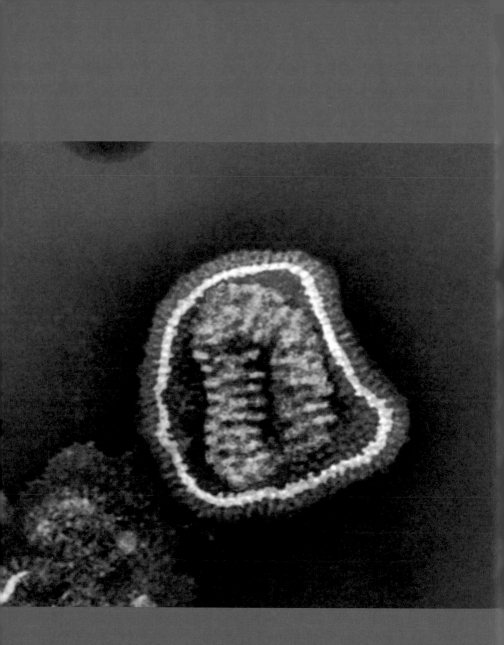

Influenza virus: the envelope layer appears orange and caspid is gray-white, with purple RNA segments inside

Looking Down from the Stars

Influenza Virus

Influenza. If you close your eyes and say the word aloud, it sounds lovely. It would make a good name for a pleasant, ancient Italian village. *Influenza* is, in fact, Italian (it means influence). It is also, in fact, an ancient name, dating back to the Middle Ages. But the charming connotations stop there. Medieval physicians believed that stars influenced the health of their patients, sometimes causing a mysterious fever that swept across Europe every few decades. And ever since, influenza has raged through our species. In 1918, a particularly virulent outbreak of the flu killed an estimated fifty million people. Even in years without an epidemic, influenza takes a brutal toll.

Each winter, thirty-six thousand people die of the flu in the United States alone; somewhere between a quarter million and a half million people die worldwide. Today scientists know that influenza is not the work of the heavens, but of a microscopic virus. Like cold-causing rhinoviruses, influenza viruses manage to wreak their harm with just ten genes. They spread in the droplets sick people release with their coughs, sneezes, and running noses. A new victim may accidentally breathe in a virus-laden droplet or pick it up on a doorknob and then bring now-contaminated fingers in contact with their mouth. Once a flu virus gets into the nose or throat, it can latch onto a cell lining the airway and slip inside. As flu viruses spread from cell to cell in the lining of the airway, they leave destruction in their wake. The mucus and cells lining the airway get destroyed, as if the flu viruses were a lawn mower cutting grass.

In healthy people, the immune system is able to launch a counterattack in a matter of days. In such cases, the flu causes a wave of aches, fevers, and fatigue, but the worst of it is over within a week. In a small fraction of its victims, the flu virus opens the way for more serious infections. Normally, the top layer of cells serves as a barrier against a wide array of pathogens. The pathogens get trapped in the mucus, and the cells snag them with hairs, swiftly notifying the immune system of intruders. Once the influenza lawnmower has cut away that protective layer, pathogens can slip in and cause dangerous lung infections, some of which can be fatal.

For a virus that has caused so much death in the past, and which continues to claim so many victims each year, influenza virus remains surprisingly mysterious. Seasonal flu is most dangerous for people with weak immune systems that can't keep the virus in check—particularly young children and the elderly. But in flu pandemics, like the 1918 outbreak, people with strong immune systems proved to be particularly vulnerable. Scientists don't know why the flu switches targets this way. One theory holds that certain strains of the flu provoke the immune system to respond so aggressively that it ends up devastating the host instead of wiping out the virus. But some scientists doubt this explanation and think the true answer lies elsewhere. Scientists also don't know when influenza viruses first started making people sick. There certainly are historical records of outbreaks of deadly fevers going back thousands of years, but it's impossible to know whether influenza viruses caused them, or another species of virus with similar symptoms.

Amidst all the mysteries of the flu, the origin of the virus is clear. It came from birds. Birds carry all known strains of human influenza viruses, along with a vast diversity of other flu viruses that don't infect humans. Many birds carry the flu without getting sick. Rather than infecting their airways, flu viruses typically infect the guts of birds; the viruses are then shed in bird droppings. Healthy birds become infected by ingesting virus-laden water.

Sometimes strains of bird flu jump the species barrier and become human viruses. But for every successful transition, there are probably many failed crossings. Bird flu viruses are well adapted to infecting their avian hosts and reproducing quickly inside them. Those adaptations make them ill-suited to spreading among humans. Starting in 2005, for example, a strain of flu from birds called H5N1 began to sicken hundreds of people in Southeast Asia. It is much deadlier than ordinary strains of seasonal flu, and so public health workers have been tracking it closely and taking measures to halt its spread. For now, at least, H5N1 can only move from a bird to a human; it cannot move from one human to another.

Unfortunately, a poorly adapted flu virus can evolve into a well-adapted one. Flu viruses are particularly sloppy at replicating their genes, so many new viruses acquire mutations. These mutations are like random changes to the letters in the flu's recipe. Some of the mutations have no effect on viruses. Some leave them unable to reproduce. But a few mutations give flu viruses a reproductive advantage. Natural selection favors these beneficial mutations, and flu strains can become better at infecting humans as mutation after mutation accumulates. Some mutations help the virus by altering the shape of the proteins that stud the virus shell, allowing them to grab human cells more effectively. Other mutations help the flu virus cope with human body temperature, which is a few degrees cooler than that of birds.

Human influenza viruses have also adapted to a new route from host to host. In birds, the viruses travel from guts to water to guts. In people, the virus moves from airways to droplets to airways. This new route also causes the flu rise and fall with the seasons. In places like the United States, most flu cases occur during the winter. According to one hypothesis, this is because the air is dry enough in those months to allow virus-laden droplets to float in the air for hours, increasing their chances of encountering a new host. In other times of the year, the humidity causes the droplets to swell and fall to the ground.

When a flu virus hitches a ride aboard a droplet and infects a new host, it sometimes invades a cell that's already harboring another flu virus. And when two different flu viruses reproduce inside the same cell, things can get messy. The genes of a flu virus are stored on eight separate segments, and when a host cell starts manufacturing the segments from two different viruses at once, they sometimes get mixed together. The new offspring end up carrying genetic material from both viruses. This mixing, known as reassortment, is a viral version of sex. When humans have children, the parents' genes are mixed together, creating new combinations of the same two sets of DNA. Reassortment allows flu viruses to mix genes together into new combinations, as well.

As scientists get a closer look at the genes of flu viruses, they're discovering that reassortment has played a major role in the natural history of the flu. A quarter of all birds with the flu have two or more virus strains inside them at once. The viruses swap genes through reassortment, and as a result they can move easily between bird species. And sometimes, on very rare occasions, an avian influenza virus can pick up human influenza virus genes through reassortment. That can be a recipe for disaster, because the new strain that results can easily spread from person to person. And because it has never circulated among humans before, no one has any defenses that could slow the new strain's spread.

Reassortment is important for other reasons than viruses jumping the species barrier. Once bird flu viruses evolve into human pathogens, they continue to swap genes among themselves every flu season. This ongoing reassortment allows the viruses to escape destruction. The longer a flu strain circulates, the more familiar it becomes to people's immune systems, and the faster they can squelch its spread. But with some viral sex, an old flu strain can pick up less familiar genes and become harder to fight.

Humans are not the only hosts who have picked up flu viruses from birds. Horses, dogs, and several other mammals have also picked it up. And in April 2009, the world became painfully aware that flu viruses also infect pigs. An outbreak of so-called swine flu jumped from pigs to humans. It first surfaced in Mexico and soon spread over the entire planet.

The history of this particular flu strain, called Human/Swine 2009 H1N1, is a tangled tale of genetic mixing and industrialized agriculture. Pigs seem to have just the right biology for reassortment; some of their receptors can easily accept human flu viruses, while other receptors welcome bird flu. Over the past century, pig farms have grown in size and density, so that flu viruses can easily move from host to host and swap genes with other strains. The oldest known swine flu strain emerged around the same time the 1918 pandemic strain entered humans; this so-called classical strain is still making pigs sick. In the 1970s a bird flu strain evolved in Europe or Asia into a new swine flu strain. A different pig-bird mix arose in the United States. And in the late 1990s, American scientists discovered a "triple reassortant" in pigs that mixed genes from all three flu strains.

Once scientists sequenced the genes of the new Human/Swine 2009 H1N1, they realized that it was the product of two different flu viruses: the triple reassortant and a Eurasian bird-to-pig strain. By comparing the new mutations that had arisen from the viruses infecting different patients, researchers have estimated that this new virus first evolved in the fall of 2008. It circulated quietly before coming to light in the spring of 2009.

Because Human/Swine 2009 H1N1 was such a new virus, public health authorities swung quickly into action. The Mexican government essentially shut down the entire country for a time, hoping to prevent the virus from finding new hosts. As Human/Swine 2009 H1N1 turned up in other countries, their governments took actions of their own. By May 2009, it was clear that while the new virus was unusually swift, it was not significantly more dangerous than typical seasonal flu.

As I write in 2010, no one can say if the new strain will fade away, outcompeted by other flu strains, or if it will mutate into a more dangerous form, or experience even more reassortment and pick up new genes. But we are not helpless as we wait to see what evolution has in store for us. We can do things to slow the spread of the flu, such as washing our hands. And scientists are learning how to make more effective vaccines by tracking the evolution of the flu virus so they can do a better job of predicting which strains will be most dangerous in flu seasons to come. We may not have the upper hand over the flu yet, but at least we no longer have to look to the stars to defend ourselves.

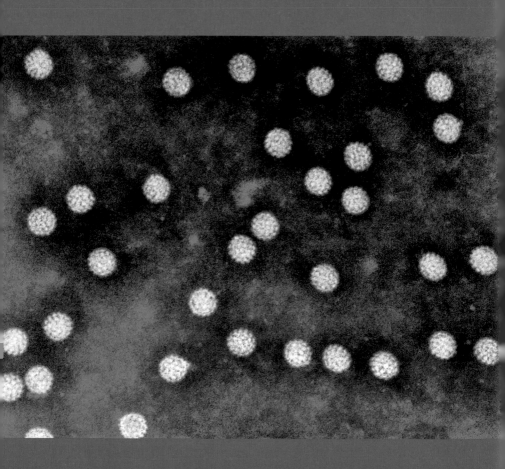

Human papillomaviruses in suspension

Rabbits with Horns

Human Papillomavirus

The stories about rabbits with horns circulated for centuries. Eventually they crystallized into the myth of the jackalope. If you go to Wyoming and twirl a rack of postcards, chances are you'll find a picture of a jackalope bounding across the prairie. It looks like a rabbit sprouting a pair of antlers. You may even see jackalopes in the flesh—or at least the head of one mounted on a diner wall.

On one level, it's all bunk. Most jackalopes are nothing but taxidermic trickery—rabbits with pieces of antelope antler glued to their heads. But like many myths, the tale of the jackalope has a grain of truth buried at its core.

Some real rabbits do indeed sprout horn-shaped growths from their heads.

In the early 1930s, Richard Shope, a scientist at Rockefeller University, heard about horned rabbits while on a hunting trip. He had a friend catch one and send him some of the tissue so that he could figure out what it was made of. Shope's colleague, Francis Rous, had done experiments with chickens that suggested viruses could cause tumors. Many scientists at the time were skeptical, but Shope wondered if rabbit "horns" were also tumors, somehow triggered by an unknown virus. To test his hypothesis, Shope ground up the horns, mixed them in a solution, and then filtered the liquid through porcelain. The fine pores of the porcelain would only let viruses through. Shope then rubbed the filtered solution onto the heads of healthy rabbits. They grew horns as well.

Shope's experiment did more than show that the horns contained viruses. He also demonstrated that the viruses *created* the horns, crafting them out of infected cells. After this discovery, Shope passed on his rabbit tissue collection to Rous, who continued to work on it for decades. Rous injected virus-loaded liquid deep inside rabbits and found that it didn't produce harmless horns. Instead, the rabbits developed aggressive cancers that killed them. For his research linking viruses and cancer, Rous won the Nobel Prize in Medicine in 1966.

The discoveries of Shope and Rous led scientists to look at growths on other animals. Cows sometimes develop monstrous lumps of deformed skin as big as grapefruits. Warts grow on mammals, from dolphins to tigers to humans. And on rare occasions, warts can turn people into human jackalopes. In the 1980s, a teenage boy in Indonesia named Dede began to develop warts on his body, and soon they had completely overgrown his hands and feet. Eventually he could no longer work at a regular job and ended up as an exhibit in a freak show, earning the nickname "Tree Man." Reports of Dede began to appear in the news, and in 2007 doctors removed thirteen pounds of warts from Dede's body. They've had to continue to perform surgeries to remove new growths from his body since then. Dede's growths, along with all the others on humans and mammals, turned out to be caused by a single virus—the same one that puts horns on rabbits. It's known as the papillomavirus, named for the papilla (*buds* in Latin) that cells form when they become infected.

In the 1970s, the German researcher Harald zur Hausen speculated that papillomaviruses might be a far bigger threat to human health than the occasional wart. He wondered whether they might also cause tumors in the cervixes of women. Previous studies on cases of cervical cancer revealed patterns that were similar to sexually transmitted diseases. Nuns, for example, get cervical cancer much less often than other women. Some scientists had speculated cervical cancer was caused by a virus spread during sex. Zur Hausen wondered if cancer-causing papillomaviruses were the culprit.

Zur Hausen reasoned that if this were true, he ought to find virus DNA in cervical tumors. He gathered biopsies to study, and slowly sorted through their DNA for years. In 1983 he discovered genetic material from papillomaviruses in the samples. As he continued to study the biopsies, he found more strains of papillomaviruses. Since zur Hausen first published his discoveries, scientists have identified one hundred different strains of human papillomavirus (or HPV for short). For his efforts, zur Hausen shared the Nobel Prize for Physiology or Medicine in 2008.

Zur Hausen's research put human papillomaviruses in medicine's spotlight, thanks to the huge toll that cervical cancer takes on the women of the world. The tumors caused by HPV grow so large that they sometimes rip the uterus or intestines apart. The bleeding can be fatal. Cervical cancer kills over 270,000 women every year, making it the third leading cause of death in women, surpassed only by breast cancer and lung cancer.

All of those cases got their start when a woman acquired an infection of HPV. The infection begins when the virus injects its DNA into a host cell. HPV specializes in infecting epithelial cells, which make up much of the skin and the body's mucous membranes. The virus's genes ends up inside the nucleus of its host cell, the home of the cell's own DNA. The cell then reads the HPV genes and makes the virus's proteins. Those proteins begin to alter the cell.

Many other viruses, such as rhinoviruses and influenza viruses, reproduce violently. They make new viruses as fast as possible, until the host cell brims with viral offspring. Ultimately, the cell rips open and dies. HPV uses a radically different strategy. Instead of killing its host cell, it causes the cell to make more copies of itself. The more host cells there are, the more viruses there are.

Speeding up a cell's division is no small feat, especially for a virus with just eight genes. The normal process of cell division is maddeningly complex. A cell "decides" to divide in response to signals both from the outside and the inside, mobilizing an army of molecules to reorganize its contents. Its internal skeleton of filaments reassembles itself, pulling apart the cell's contents to two ends. At the same time, the cell makes a new copy of its DNA—3.5 billion "letters" all told, organized into 46 clumps called chromosomes. The cell must drag those chromosomes to either end of the cell and build a wall through its center. During this buzz of activity, supervising molecules monitor the progress. If they sense that the division is going awry—if the cell acquires a defect that might make it cancerous, for example—the monitor molecules trigger the cell to commit suicide. HPV can manipulate this complex dance by producing just a few proteins that intervene at crucial points in the cell cycle, accelerating it without killing the cell.

Many types of cells grow quickly in childhood and then slow down or stop altogether. Epithelial cells, the cells that HPV infects, continue to grow through our whole life. They start out in a layer buried below the skin's surface. As they divide, they produce a layer of new cells that pushes up on the cells above them. As the cells divide and rise, they become different than their progenitors. They begin to make more of a hard protein called keratin (the same stuff that makes up fingernails and horse hooves). Loaded with keratin, the top layer of skin can better withstand the damage from the sun, chemicals, and extreme temperatures. But eventually the top layer of epithelial cells dies off, and the next rising layers of epithelial cells take its place.

This arrangement means that HPV has to try to live on a conveyor belt. As HPV-infected cells reproduce, they move upward, closer and closer to their death. The viruses sense when their host cells are getting close to the surface and shift their strategy. Instead of speeding up cell division, they issue commands to their host cell to make many new viruses. When the cell reaches the surface, it bursts open with a big supply of HPV that can seek out new hosts to infect.

For most people infected with HPV, a peaceful balance emerges between virus and host. Fast-growing infected cells don't cause people harm, because they get sloughed off. The virus, meanwhile, gets to use epithelial cells as factories for new viruses, which can then infect new hosts through skin-to-skin contact and sex. The immune system helps maintain the balance by clearing away some of the infected cells. (Dede's tree-like growths were the result of a genetic defect that left his body unable to rein in the virus.)

This balance between host and virus has existed for hundreds of millions of years. To reconstruct the history of papillomaviruses, scientists compare the genetic sequence of different strains and note which animals they infect. It turns out that papillomaviruses infect not just mammals, such as humans, rabbits, and cows, but other vertebrates as well, such as birds and reptiles. Each strain of virus typically only infects one or a few related species. Based on their relationships, Marc Gottschling of the University of Munich has argued that the first egg-laying land vertebrates— the ancestor of mammals, reptiles, and birds—was already a host to papillomaviruses three hundred million years ago.

As the descendants of that ancient animal evolved into different lineages, their papillomaviruses evolved as well. Some research suggests that these viruses began to specialize on different kinds of lining in their hosts. The viruses that cause warts, for example, adapted to infect skin cells. Another lineage adapted to the mucosal linings of the mouth and other orifices. For the most part, these new papillomaviruses coexisted peacefully with their hosts. Two-thirds of healthy horses carry strains of papillomavirus called BPV1 and BPV2. Some strains evolved to be more prone to turn cancerous than others, but researchers can't say why.

For thousands of generations, papillomaviruses would specialize on certain hosts, but from time to time, they leap to new species. A number of human papillomaviruses are most closely related to papillomaviruses that infect distantly related animals, like horses, instead of our closest ape relatives. Nothing more than skin contact may have been enough to allow viruses to make the jump.

When our own species first evolved in Africa about two hundred thousand years ago, our ancestors probably carried several different strains of papillomaviruses. Representatives of those strains can be found all over the world. But as humans expanded across the planet—leaving Africa about fifty thousand years ago and reaching the New World by about fifteen thousand years ago—their papillomaviruses were continuing to evolve. We know this because the genealogy of some HPV strains reflect the genealogy of our species. The viruses that infect living Africans belong to the oldest lineages of HPV, for example, while Europeans, Asians, and Native Americans carry their own distinct strains.

For about 199,950 of the past 200,000 years, our species had no idea that we were carrying HPV. That's not because HPV was a rare virus—far from it: a 2008 study on 1,797 men and women found 60 percent of them had antibodies to HPV, indicating they had been infected with the virus at some point in their life. For the overwhelming majority of those people, the experience was harmless. Of the estimated 30 million American women who carry HPV, only 13,000 a year develop cervical cancer.

In this cancer-stricken minority, the peaceful balance between host and virus is thrown off. Each time an infected cell divides, there's a small chance it will mutate one of the genes that helps regulate the cell cycle. In an uninfected cell, the mutation would not do much harm. But a cell that's already being pushed by HPV to grow faster is in a precarious state. What might otherwise be a harmless mutation transforms an infected cell into a precancerous one. The cell multiplies much faster than before. Its descendants grow so fast that the shedding of the top layer of epithelial cells is not enough to get rid of them. They form a tumor, which pushes out and down into the surrounding tissue.

The best way to prevent most cancers is to reduce the odds that our cells will pick up dangerous mutations: quitting smoking, avoiding cancer-promoting chemicals, and eating well. But cervical cancer can be blocked another way: with a vaccine. In 2006, the first HPV vaccines were approved for use in the United States and Europe. They all contain proteins from the outer shell of HPV, which the immune system can learn to recognize. If people are later infected with HPV, their immune system can mount a rapid attack and wipe it out.

The introduction of the vaccines has brought controversies of many flavors. The developers recommend the vaccines for girls in their early teens. Some parents have protested that such a policy promotes sex before marriage. In 2008, medical experts raised a different set of concerns in editorials in the *New England Journal of Medicine*. It takes many years for HPV to give rise to cancers, they pointed out, and so we don't yet know how effective the vaccines will prove to be.

Another potential problem is the fact that current HPV vaccines only target two strains of the virus. The choice makes a certain amount of sense for vaccine makers who have to balance costs and benefits, since those two strains cause about 70 percent of all cases of cervical cancer. But we humans are host to over a hundred different strains of HPV, which are constantly acquiring new mutations and swapping genes between one another. If vaccines decimate the two most successful strains, natural selection might well favor the evolution of other strains to take their place. Never underestimate the evolutionary creativity of a virus that can transform rabbits into jackalopes or men into trees.

EVERYWHERE,

IN ALL THINGS

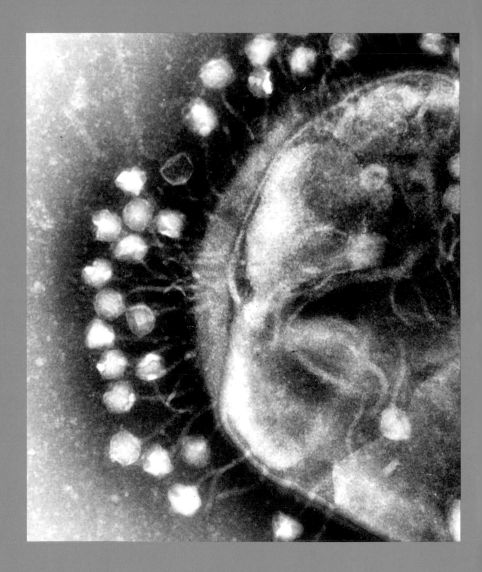

Bacteriophages attach to the surface of the host cell,
the bacterium *Escherichia coli*

The Enemy of Our Enemy

Bacteriophages

People have known about viruses, or at least their effects, for as long as viruses have been making people sick. Scientists discovered viruses in the nineteenth century, and by the beginning of the twentieth, they had learned a few important things about them. They knew that viruses were infectious agents of unimaginably small size. They had begun to assign certain diseases, such as tobacco mosaic disease or rabies, to certain viruses. But the young science of virology was still parochial. It focused mainly on the viruses that worried people most: the ones that infect humans or the ones that infect the crops and livestock we raise for food. Virologists rarely looked beyond our little circle of experience.

A clue to the true scope of viruses came in the middle of World War I. French soldiers were dying in droves, killed not just by Germans but also by bacteria. The microbes invaded their torn flesh, their food, and their drinking water. Their path was made easier by the worldwide flu epidemic in 1918. The flu weakened the defenses of its victims, allowing bacteria to infect their lungs. The soldiers spread the flu to civilians, and ultimately fifty million people died—many of them killed by bacteria.

Today, doctors can treat all of these bacterial infections with antibiotics. But antibiotics would not be discovered until the 1930s. During World War I, doctors could only treat battlefield infections by cleaning wounds and, if that failed, amputating limbs. Their patients often died anyway.

In 1917, in the midst of this carnage, the Canadian-born physician Felix d'Herelle discovered what seemed to him a medical miracle: a powerful substance that could wipe out bacteria. It was not an antibiotic. Instead, Herelle had discovered something that no one had ever imagined before: a virus that attacked not humans, or other animals, or even plants. He found a virus that made bacteria its host.

Herelle made his discovery while investigating an outbreak of dysentery among French soldiers. As part of his analysis, he passed the stool of the soldiers through a filter. The filter's pores were so small that not even the bacteria that caused the dysentery, known as *Shigella*, could slip through. Once Herelle had produced this clear, filtered fluid, he then mixed it with a fresh sample of *Shigella* bacteria and then spread the mixture of bacteria and clear fluid in petri dishes.

The *Shigella* began to grow, but within a few hours Herelle noticed strange clear spots starting to form in their colonies. He drew samples from those spots and mixed them with *Shigella* again. More clear spots formed in the dishes. These spots, Herelle concluded, were bacteria battlegrounds in which viruses were killing *Shigella* and leaving behind their translucent corpses. Herelle believed his discovery was so radical that his viruses deserved a name of their own. He dubbed them *bacteriophages*, meaning "eaters of bacteria." Today, they're known as phages for short.

The concept of bacteria-infecting viruses was so strange and so new that some scientists couldn't believe it. Jules Bordet, a French immunologist who won the Nobel Prize in 1919, became Herelle's most outspoken critic after he failed to find phages of his own. Instead of *Shigella*, Bordet used a harmless strain of *Escherichia coli*. He poured *E. coli*–laden liquid through fine filters, and then mixed the filtered liquid with a second batch of *E. coli*. The second batch died, just as they had in Herelle's experiments. But then Bordet decided to see what would happen if he mixed the filtered liquid with the first batch of *E. coli*—that is, the one he had filtered in the first place. To his surprise, the first batch of *E. coli* was immune. Bordet believed that his failure to kill the bacteria meant that the filtered fluid did not contain phages. Instead, he thought, it contained a protein produced by the first *E. coli*. The protein was toxic to other bacteria, but not to the ones that made it.

Herelle fought back, Bordet counterattacked, and the debate raged for years. It wasn't until the 1940s that scientists finally found the visual proof that Herelle was right. By then, engineers had built electron microscopes powerful enough to let scientists see viruses. When they mixed bacteria-killing fluid with *E. coli* and put it under the microscopes, they saw that bacteria were attacked by phages. The phages had boxlike shells in which their genes were coiled, sitting atop a set of what looked like spider legs. The phages dropped onto the surface of *E. coli* like a lunar lander on the moon and then drilled into the microbe, squirting in their DNA.

As scientists got to know phages better, it became clear that the debate between Herelle and Bordet was just a case of apples and oranges. Phages do not belong to a single species, and different phage species behave differently toward their hosts. Herelle had found a vicious form, called a lytic phage, which kills its host as it multiplies. Bordet had found a more benevolent kind of virus, which came to be known as a temperate phage. Temperate phages treat bacteria much like human papillomaviruses treat our skin cells. When a temperate phage infects its host microbe, its host does not burst open with new phages. Instead, the temperate phage's genes are joined into the host's own DNA, and the host continues to grow and divide. It is as if the virus and its host become one.

Once in a while, however, the DNA of the temperate phage awakens. It commandeers the cell to make new phages, which burst out of the cell and invade new ones. And once a temperate phage is incorporated into a microbe, the host becomes immune from any further invasion. That's why Bordet couldn't kill his first batch of E. coli with the phage—it was already infected, and thus protected.

Herelle did not wait for the debate over phages to end before he began to use them to cure his patients. During World War I, he observed that as soldiers recovered from dysentery and other diseases, the levels of phages in their stool climbed. Herelle concluded that the phages were actually killing the bacteria. Perhaps, if he gave his patients extra phages, he could eliminate diseases even faster.

Before he could test this hypothesis, Herelle first needed to be sure phages were safe. So he swallowed some to see if they made him sick. He found that he could ingest phages, as he later wrote, "without detecting the slightest malaise." Herelle injected phages into his skin, again with no ill effects. Confident that phages were safe, Herelle began to give them to sick patients. He reported that they helped people recover from dysentery and cholera. When four passengers on a French ship in the Suez Canal came down with bubonic plague, Herelle gave them phages. All four victims recovered.

Herelle's cures made him even more famous than before. The American writer Sinclair Lewis made Herelle's radical research the basis of his 1925 best-selling novel *Arrowsmith*, which Hollywood turned into a movie in 1931. Meanwhile, Herelle developed phage-based drugs sold by the company that's now known as L'Oreal. People used his phages to treat skin wounds and to cure intestinal infections.

But by 1940, the phage craze had come to end. The idea of using live viruses as medicine had made many doctors uneasy. When antibiotics were discovered in the 1930s, those doctors responded far more enthusiastically, because antibiotics were not alive; they were just artificial chemicals and proteins produced by fungi and bacteria. Antibiotics were also staggeringly effective, often clearing infections in a few days. Pharmaceutical companies abandoned Herelle's phages and began to churn out antibiotics. With the success of antibiotics, investigating phage therapy seemed hardly worth the effort.

Yet Herelle's dream did not vanish entirely when he died in 1949. On a trip to the Soviet Union in the 1920s, he had met scientists who wanted to set up an entire institute for research on phage therapy. In 1923 he helped Soviet researchers establish the Eliava Institute of Bacteriophage, Microbiology, and Virology in Tbilisi, which is now the capital of the Republic of Georgia. At its peak, the institute employed 1,200 people to produce tons of phages a year. During World War II, the Soviet Union shipped phage powders and pills to the front lines, where they were dispensed to infected soldiers.

In 1963, the Eliava Institute ran the largest trial ever conducted to see how well phages actually worked in humans, enrolling 30,769 children in Tbilisi. Once a week, about half the children swallowed a pill that contained phages against Shigella. The other half of the children got a pill made of sugar. To minimize environmental influences as much as possible, the Eliava scientists gave the phage pills only to children who lived on one side of each street, and the sugar pills to the children who lived on the other side. The scientists followed the children for 109 days. Among the children who took the sugar pill, 6.7 out of every 1,000 got dysentery. Among the children who took the phage pill, that figure dropped to 1.8 per 1,000. In other words, taking phages caused a 3.8-fold decrease in a child's chance of getting sick.

Few people outside of Georgia heard about these striking results, thanks to the secrecy of the Soviet government. Only after the Soviet Union fell in 1989 did news start to trickle out. The reports have inspired a small but dedicated group of Western scientists to investigate phage therapy and to challenge the long-entrenched reluctance in the West to use them.

These phage champions argue that we should not be worried about using live viruses as medical treatments. After all, phages swarm inside many of the foods we eat, such as yogurt, pickles, and salami. Our bodies are packed with phages too, which is not surprising when you consider that we each carry about a hundred trillion bacteria—all promising hosts for various species of phages. Every day, those phages kill vast numbers of bacteria inside our bodies without ever harming our health.

Another concern that's been raised about phages is that their attack is too narrowly focused. Each species of phage can only attack one species of bacteria, while one antibiotic can kill off many different species at once. But it's clear now that phage therapy can treat a wide range of infections. Doctors just have to combine many phage species into a single cocktail. Scientists at the Eliava Institute have developed a dressing for wounds that is impregnated with half a dozen different phages, capable of killing the six most common kinds of bacteria that infect skin wounds.

Skeptics have also argued that even if scientists could design an effective phage therapy, evolution would soon render it useless. In the 1940s, the microbiologists Salvador Luria and Max Delbruck observed phage resistance evolving before their own eyes. When they laced a dish of E. coli with phages, most of the bacteria died, but a few clung to existence and then later multiplied into new colonies. Further research revealed that those survivors had acquired mutations that allowed them to resist the phages. The resistant bacteria then passed on their mutated genes to their descendants. Critics have argued that phage therapy would also foster the evolution of phage-resistant bacteria, allowing infections to rebound.

The advocates for phage therapy respond by pointing out that phages can evolve, too. As they replicate, they sometimes pick up mutations, and some of those mutations can give them new avenues for infecting resistant bacteria. Scientists can even help phages improve their attacks. They can search through collections of thousands of different phages to find the best weapon for any particular infection, for example. They can even tinker with phage DNA to create phages that can kill in new ways.

In 2008, James Collins, a biologist at Boston University, and Tim Lu of MIT published details of the first phage engineered to kill. Their new phage is especially effective because it's tailored to attack the rubbery sheets that bacteria embed themselves in, known as biofilms. Biofilm can foil antibiotics and phages alike, because they can't penetrate the tough goo and reach the bacteria inside. Collins and Lu searched through the scientific literature for a gene that might make phages better able to destroy biofilms. Bacteria themselves carry enzymes that they use to loosen up biofilms when it's time for them to break free and float away to colonize new habitats. So Collins and Lu synthesized a gene for one of these biofilm-dissolving enzymes and inserted it into a phage. They then tweaked the phage's DNA so that it would produce lots of the enzyme as soon as it entered a host microbe. When they unleashed it on biofilms of *E. coli*, the phages penetrated the microbes on the top of the biofilms and forced them to make both new phages and new enzymes. The infected microbes burst open, releasing enzymes that sliced open deeper layers of the biofilms, which the phages could infect. The engineered phages can wipe 99.997 percent of the *E. coli* in a biofilm, a kill rate that's about a hundred times better than ordinary phages.

While Collins and other scientists discover how to make phages even more effective, antibiotics are now losing their luster. Doctors are grappling with a growing number of bacteria that have evolved resistance to most of the antibiotics available today. Sometimes doctors have to rely on expensive, last-resort drugs that come with harsh side effects. And there's every reason to expect that bacteria will evolve to resist last-resort antibiotics as well. Scientists are scrambling to develop new antibiotics, but it can take over a decade to get a new drug from the lab to the marketplace. It may be hard to imagine a world before antibiotics, but now we must imagine a world where antibiotics are not the only weapon we use against bacteria. And now, ninety years after Herelle first encountered bacteriophages, these viruses may finally be ready to become a part of modern medicine.

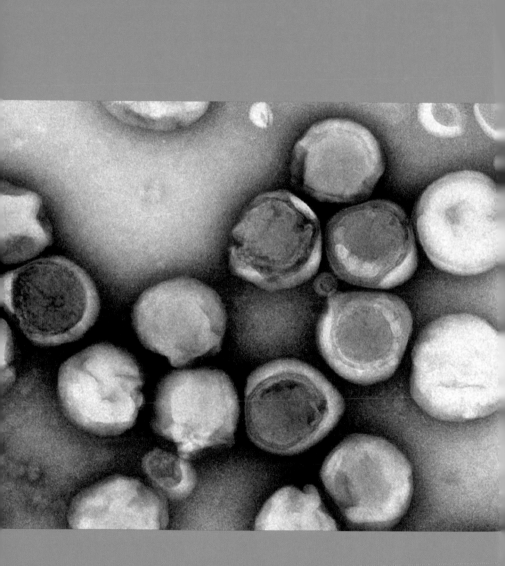

Emiliania huxleyi viruses infect ocean algae
(the viruses shown here in suspension)

The Infected Ocean

Marine Phages

Some great discoveries seem at first like terrible mistakes.

In 1986 a graduate student at the State University of New York at Stony Brook named Lita Proctor decided to see how many viruses there are in seawater. At the time, the general consensus was that there were hardly any. The few researchers who had bothered to look for viruses in the ocean had generally found only a scarce supply. Most experts believed that the majority of the viruses they did find in sea water had actually come from sewage and other sources on land.

But over the years, a handful of scientists had gathered evidence that didn't fit neatly into the consensus. A marine biologist named John Sieburth had published a photograph of a marine bacterium erupting with new viruses, for example. Proctor decided it was time to launch a systematic search. She traveled to the Caribbean and to the Sargasso Sea, scooping up seawater along the way. Back on Long Island, she carefully extracted the biological material from the seawater, which she coated with metal so that it would show up under the beam of an electron microscope. When Procter finally looked at her samples, she beheld a world of viruses. Some floated freely, while others were lurking inside infected bacterial hosts. Based on the number of viruses she found in her samples, Proctor estimated that every liter of seawater contained up to one hundred billion viruses.

Proctor's figure was far beyond anything that had come before. It would have surprised few scientists if she had turned out to have added on a few extra zeroes by accident. But when other scientists carried out their own surveys, they ended up with similar estimates. Scientists came to agree that there are somewhere in the neighborhood of 1,000,000,000,000,000,000,000,000,000,000 viruses in the ocean.

It is hard to find a point of comparison to make sense of such a huge number. Viruses outnumber all other residents of the ocean by about fifteen to one. If you put all the viruses of the oceans on a scale, they would equal the weight of seventy-five million blue whales. And if you lined up all the viruses in the ocean end to end, they would stretch out past the nearest sixty galaxies.

These numbers don't mean that a swim in the ocean is a death sentence. Only a minute fraction of the viruses in the ocean can infect humans. Some marine viruses infect fishes and other marine animals, but by far their most common targets are microbes. Microbes may be invisible to the naked eye, but collectively they dwarf all the ocean's whales, its coral reefs, and all other forms of marine life. And just as the bacteria that live in our bodies are attacked by phages, marine microbes are attacked by marine phages.

When Felix d'Herelle discovered the first bacteriophage in French soldiers in 1917, many scientists refused to believe that such a thing actually existed. A century later, it's clear that Herelle had found the most abundant life form on Earth. Ever since Proctor's discovery of the abundance of marine viruses, scientists have been documenting their massive influence on the planet. Marine phages influence the ecology of the world's oceans. They leave their mark on Earth's global climate. And they have been playing a crucial part in the evolution of life for billions of years. They are, in other words, biology's living matrix.

Marine viruses are powerful because they are so infectious. They invade a new microbe host ten trillion times a second, and every day they kill about half of all bacteria in the world's oceans. Their lethal efficiency keeps their hosts in check, and we humans often benefit from their deadliness. Cholera, for example, is caused by blooms of waterborne bacteria called *Vibrio*. But *Vibrio* are host to a number of phages. When the population of *Vibrio* explodes and causes a cholera epidemic, the phages multiply. The virus population rises so quickly that it kills *Vibrio* faster than the microbes can reproduce. The bacterial boom subsides, and the cholera epidemic fades away.

Stopping cholera outbreaks is actually one of the smaller effects of marine viruses. They kill so many microbes that they can also influence the atmosphere across the planet. That's because microbes themselves are the planet's great geoengineers. Algae and photosynthetic bacteria churn out about half of the oxygen we breathe. Algae also release a gas called dimethyl sulfide that rises into the air and seeds clouds. The clouds reflect incoming sunlight back out into space, cooling the planet. Microbes also absorb and release vast amounts of carbon dioxide, which traps heat in the atmosphere. Some microbes release carbon dioxide into the atmosphere as waste, warming the planet. Algae and photosynthetic bacteria, on the other hand, suck carbon dioxide in as they grow, making the atmosphere cooler. When microbes in the ocean die, some of their carbon rains down to the sea floor. Over millions of years, this microbial snow can steadily make the planet cooler and cooler. What's more, these dead organisms can turn to rock. The White Cliffs of Dover, for example, are made up of the chalky shells of single-cell organisms called coccolithophores.

Viruses kill these geoengineers by the trillions every day. As their microbial victims die, they spill open and release a billion tons of carbon a day. Some of the liberated carbon acts as a fertilizer, stimulating the growth of other microbes, but some of it probably sinks to the bottom of the ocean. The molecules inside a cell are sticky, and so once a virus rips open a host, the sticky molecules that fall out may snag other carbon molecules and drag them down in a vast storm of underwater snow.

Ocean viruses are stunning not just for their sheer numbers but also for their genetic diversity. The genes in a human and the genes in a shark are quite similar—so similar that scientists can find a related counterpart in the shark genome to most genes in the human genome. The genetic makeup of marine viruses, on the other hand, matches almost nothing. In a survey of viruses in the Arctic Ocean, the Gulf of Mexico, Bermuda, and the northern Pacific, scientists identified 1.8 million viral genes. Only 10 percent of them showed any match to any gene from any microbe, animal, plant, or other organism—even from any other known virus. The other 90 percent were entirely new to science. In 200 liters of seawater, scientists typically find 5,000 genetically distinct kinds of viruses. In a kilogram of marine sediment, there may be a million kinds.

One reason for all this diversity is that marine viruses have so many hosts to infect. Each lineage of viruses has to evolve new adaptations to get past its host's defenses. But diversity can also evolve by more peaceful means. Temperate phages merge seamlessly into their host's DNA; when the host reproduces, it copies the virus's DNA along with its own. As long as a temperate phage's DNA remains intact, it can still break free from its host during times of stress. But over enough generations, a temperate phage will pick up mutations that hobble it, so that it can no longer escape. It becomes a permanent part of its host's genome.

As a host cell manufactures new viruses, it sometimes accidentally adds some of its own genes to them. The new viruses carry the genes of their hosts as they swim through the ocean, and they insert them, along with their own, into the genomes of their new hosts. By one estimate, viruses transfer a trillion trillion genes between host genomes in the ocean every year.

Sometimes these borrowed genes make the new host more successful at growing and reproducing. The success of the host means success for the virus, too. While some species of viruses kill *Vibrio*, others deliver genes for toxins that the bacteria use to trigger diarrhea during cholera infections. The new infection of toxin-carrying viruses may be responsible for new cholera outbreaks.

Thanks to gene borrowing, viruses may also be directly responsible for a lot of the world's oxygen. An abundant species of ocean bacteria, called *Synechococcus*, carries out about a quarter of the world's photosynthesis. When scientists examine the DNA of *Synechococcus* samples, they often find proteins from viruses carrying out their light harvesting. Scientists have even found free-floating viruses with photosynthesis genes, searching for a new host to infect. By one rough calculation, 10 percent of all the photosynthesis on Earth is carried out with virus genes. Breathe ten times, and one of those breaths comes to you courtesy of a virus.

This shuttling of genes has had a huge impact on the history of all life on Earth. It was in the oceans that life got its start, after all. The oldest traces of life are fossils of marine microbes dating back almost 3.5 billion years. It was in the oceans that multicellular organisms evolved; their oldest fossils date back to about 2 billion years ago. In fact, our own ancestors did not crawl onto land until about 400 million years ago. Viruses don't leave behind fossils in rocks, but they do leave marks on the genomes of their hosts. Those marks suggest that viruses have been around for billions of years.

Scientists can determine the history of genes by comparing the genomes of species that split from a common ancestor that lived long ago. Those comparisons can, for example, reveal genes that were delivered to their current host by a virus that lived in the distant past. Scientists have found that all living things have mosaics of genomes, with hundreds or thousands of genes imported by viruses. As far down as scientists can reach on the tree of life, viruses have been shuttling genes. Darwin may have envisioned the history of life as a tree. But the history of genes, at least among the ocean's microbes and their viruses, is more like a bustling trade network, its webs reaching back billions of years.

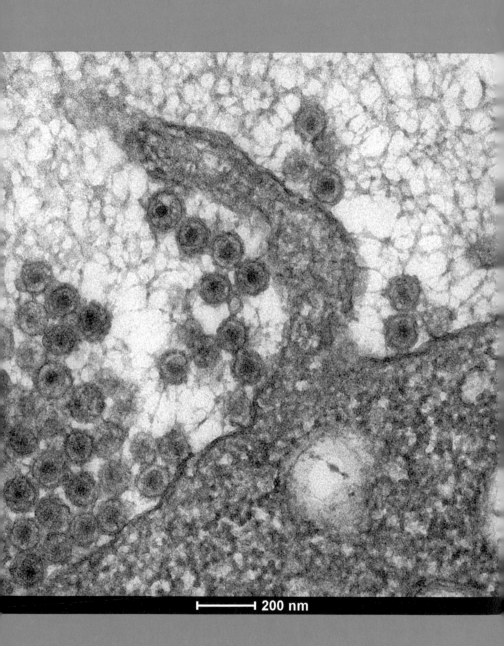

Avian leukocyte viruses bud from a human white blood cell

Our Inner Parasites

Endogenous Retroviruses

The idea that a host's genes could have come from viruses is almost philosophical in its weirdness. We like to think of genomes as our ultimate identity. We know who our biological parents are because they gave us our DNA. In our DNA are not just the instructions for the color of our skin or our susceptibility to diabetes. Our very nature lurks there. That's why the idea of cloning is so abhorrent: no one should have to carry secondhand genes.

But if most of an organism's genes arrived in its genome in a virus, does it even have a distinct identity of its own? Or is it just a mishmash of genes, cobbled together by evolution? It's as if the world was filled with hybrid monsters, with clear lines of identity blurred away.

Microbiologists have been getting used to the viral roots of the microbes they study for decades now. And as long as microbes were the only organisms with much evidence of virus-imported genes, we could try to ignore this philosophical weirdness by thinking of it merely as a fluke of "lower" life forms. But now we can no longer find comfort this way. If we look inside our own genome, we now see viruses. Thousands of them.

We have the jackalope to thank for this realization. The myth of the jackalope was one of the clues that led virologists to discover that some viruses cause cancer. In the 1960s, one of the most intensely studied cancer-causing viruses was avian leukosis virus. At the time, the virus was sweeping across chicken farms and threatening the entire poultry industry. Scientists found that avian leukosis virus belonged to a group of species known as retroviruses. Retroviruses insert their genetic material into their host cell's DNA. When the host cell divides, it copies the virus's DNA along with its own. Under the certain conditions, the cell is forced to produce new viruses—complete with genes and a protein shell—which can then escape to infect a new cell. Retroviruses sometimes trigger cells to turn cancerous if their genetic material is accidentally inserted in the wrong place in their host's genome. Retroviruses have genetic "on switches" that prompt their host cell to make proteins out of nearby genes. Sometimes their switches turn on host genes that ought to be kept shut off, and cancer can result.

Avian leukosis virus proved to be a very strange retrovirus. At the time, scientists tested for the presence of the virus by screening chicken blood for one of the virus's proteins. Sometimes they would find the avian leukosis virus protein in the blood of chickens that were perfectly healthy and never developed cancer. Stranger still, healthy hens carrying the protein could produce chicks that were also healthy and also carried the protein.

Robin Weiss, a virologist then working at the University of Washington, wondered if the virus had become a permanent, harmless part of the chicken DNA. He and his colleagues treated cells from healthy chickens with mutation-triggering chemicals and radiation to see if they could flush the virus out from its hiding place. Just as they had suspected, the mutant cells started to churn out the avian leukosis virus. In other words, these healthy chickens were not simply infected with avian leukosis virus in some of their cells; the genetic instructions for making the virus were implanted in *all* of their cells, and they passed those instructions down to their descendants.

These hidden viruses were not limited to just one oddball breed of chickens. Weiss and other scientists found avian leukosis virus embedded in many breeds, raising the possibility that the virus was an ancient component of chicken DNA. To see just how long ago avian leukosis viruses infected the ancestors of today's chickens, Weiss and his colleagues travelled to the jungles of Malaysia. There they trapped red jungle fowl, the closest wild relatives of chickens. The red jungle fowl carried the same avian leukosis virus, Weiss found. On later expeditions, he found that other species of jungle fowl lacked the virus.

Out of the research on avian leukosis virus emerged a hypothesis for how it had merged with chickens. Thousands of years ago, the virus plagued the common ancestor of domesticated chickens and red jungle fowl. It invaded cells, made new copies of itself, and infected new birds, leaving tumors in its wake. But in at least one bird, something else happened. Instead of giving the bird cancer, the virus was kept in check by the bird's immune system. As it spread harmlessly through the bird's body, it infected the chicken's sexual organs. When an infected bird mated, its fertilized egg also contained the virus's DNA in its own genes.

As the infected embryo grew and divided, all of its cells also inherited the virus DNA. When the chick emerged from its shell, it was part chicken and part virus. And with the avian leukosis virus now part of its genome, it passed down the virus's DNA to its own offspring. The virus remained a silent passenger from generation to generation for thousands of years. But under certain conditions, the virus could reactivate, create tumors, and spread to other birds.

Scientists recognized that this new virus was in a class of its own. They called it an endogenous retrovirus—endogenous meaning *generated within*. They soon found endogenous retroviruses in other animals. In fact, the viruses lurk in the genomes of just about every major group of vertebrates, from fish to reptiles to mammals. Some of the new endogenous retroviruses turned out to cause cancer like avian leukosis virus, but many did not. Some seemed to be effectively muzzled by their host. Certain endogenous retroviruses carried by mice cannot infect mice cells, for example, but they can readily spread among rat cells.

Other endogenous retroviruses turned out to be crippled, carrying mutations that robbed them of the ability to make full-fledged viruses. They could still make new copies of their genes, however, which were then reinserted back into their host's genome. And scientists also discovered some endogenous retroviruses that were so riddled with mutations that they could no longer do anything at all. They had become nothing more than baggage in their host's genome.

Endogenous retroviruses can linger in their hosts for millions of years. In 2009, Aris Katzourakis, an evolutionary biologist at the University of Oxford, discovered hundreds of copies of endogenous retroviruses in the genome of the three-toed sloth. Their genes closely matched those of foamy viruses, free-living pathogens that infect primates and other mammals. Katzourakis concluded that foamy viruses infected the common ancestor of three-toed sloths and primates, which lived a hundred million years ago. In primates, they've remained free-living. In the sloth lineage, however, they became trapped in their host's DNA and have remained there ever since.

As scientists discovered endogenous retroviruses in other species, they naturally wondered about our own DNA. After all, we suffer infections from many retroviruses. Virologists tried coaxing endogenous retroviruses out of human cells without any luck. But when they scanned the human genome, they found many segments of DNA that bore a striking resemblance to retroviruses. Many of those segments resembled retrovirus-like segments in apes and monkeys, suggesting that they had infected our ancestors thirty million years ago or more. But some of the retrovirus-like segments in the human genome had no counterparts in any other species. It was possible that the segments unique to humans started out as retroviruses that infected our ancestors a million years ago.

To test this idea, Thierry Heidmann, a researcher at the Gustave Roussy Institute in Villejuif, France, tried to bring a human endogenous retrovirus back to life. Searching through the genomes of different people, he and his colleagues found slightly different versions of one retrovirus-like segment. These differences presumably arose after a retrovirus became trapped in the genomes of ancient humans. In their descendants, mutations struck different parts of the virus's DNA.

Heidmann and his colleagues compared the variants of the virus-like sequence. It was as if they found four copies of a play by Shakespeare, each transcribed by a slightly careless clerk. Each clerk might make his own set of mistakes. Each copy might have a different version of the same word—say, *wheregore*, *sherefore*, *whorefore*, *wherefrom*. By comparing all four versions, an historian could figure out that the original word was *wherefore*.

Using this method, Heidmann and his fellow scientists were able to use the mutated versions in living humans to determine the original sequence of the DNA. They then synthesized a piece of DNA with a matching sequence and insert it into human cells they reared in a culture dish. Some of the cells produced new viruses that could infect other cells. In other words, the original sequence of the DNA had been a living, functioning virus. In 2006, Heidmann named the virus Phoenix, for the mythical bird that rose from its own ashes.

Retroviruses are a major threat to human health when they're free-living, but even after they become endogenous they remain dangerous. Mutations can give them back the ability to make full-blown viruses that can escape and cause new infections and even cause cancer. Endogenous retroviruses that can only insert new copies of their DNA into their host genome are dangerous as well, because they can cause genes that are shut down to switch on at the wrong times. The threat from endogenous retroviruses is so great, in fact, that our ancestors evolved weapons that exist only to keep these viruses from spreading.

Paul Bieniasz, a virologist at Rockefeller University, discovered two of these weapons in 2007 by reviving an endogenous retrovirus, as Hiedmann's team had revived Phoenix the year before. Bieniasz dubbed his resurrected virus HERV-K[con]. When he infected human cells with it, he found that the cells could fight the virus using two proteins called APOBEC3. Bieniasz's experiments suggest that APOBEC3 homes in on endogenous retroviruses as they are making new copies of themselves destined to be inserted back into the host's genome. The protein upsets the gene-copying process so that the new copies of the viruses pick up extra mutations. The extra mutations act like a hail of bullets. Some of them don't cause any harm, but if one of them hits a vital spot in the virus's DNA, it can cripple the virus so that it can no longer reproduce.

Proteins like APOBEC3 disable endogenous retroviruses, but they don't eliminate them. Over millions of years, our genomes have picked up a vast amount of DNA from dead viruses. Each of us carries almost a hundred thousand fragments of endogenous retrovirus DNA in our genome, making up about 8 percent of our DNA. To put that figure in perspective, consider that the twenty thousand protein-coding genes in the human genome make up only 1.2 percent of our DNA. Scientists have also observed millions of smaller pieces of "jumping DNA" in the human genome. It's possible that many of those pieces evolved from endogenous retrovirus, having been stripped down to the bare essentials required for copying DNA.

Endogenous retroviruses may be dangerous parasites, but scientists have discovered a few that we have commandeered for our own benefit. When a fertilized egg develops into a fetus, for example, some of its cells develop into the placenta, an organ that draws in nutrients from the mother's tissues. The cells in the outer layer of the placenta fuse together, sharing their DNA and other molecules. Heidmann and other researchers have found that a human endogenous retrovirus gene plays a crucial role in that fusion. The cells in the outer placenta use the gene to produce a protein on their surface, which latches them to neighboring cells. In our most intimate moment, as new human life emerges from old, viruses are essential to our survival. There is no us and them—just a gradually blending and shifting mix of DNA.

THE VIRAL FUTURE

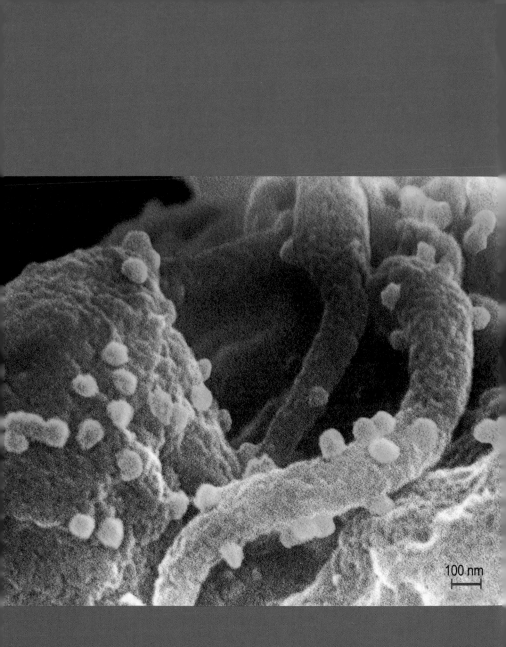

100 nm

Human immunodeficiency viruses on the surface of a CD4 white blood cell

The Young Scourge

Human Immunodeficiency Virus

Every week, the Centers for Disease Control and Prevention publish a thin newsletter called *Morbidity and Mortality Weekly Report*. The issue that appeared on July 4, 1981, was a typical assortment of the ordinary and the mysterious. Among the mysteries that week was a report from Los Angeles, where doctors had noticed an odd coincidence. Between October 1980 and May 1981, five men were admitted to hospitals around the city with the same rare disease, known as pneumocystis pneumonia.

Pneumocystis pneumonia is caused by a common fungus called *Pneumocystis jiroveci*. The spores of *P. jiroveci* are so abundant that most people inhale it at some point during their childhood. Their immune system quickly kills off the fungus and produces antibodies that ward off any future infection. But in people with weak immune systems, *P. jiroveci* runs rampant. The lungs fill with fluid and become badly scarred. Its victims struggle to breathe enough oxygen to stay alive. The five Los Angeles patients did not fit the typical profile of a pneumocystis pneumonia victim. They were young men who had been in perfect health before they came down with pneumonia. Commenting on the report, the editors of *Morbidity and Mortality Weekly Report* speculated that the puzzling symptoms of the five men "suggest the possibility of a cellular-immune dysfunction."

Little did they know that they were publishing the first observations of what would become the greatest epidemic in modern history. The five Los Angeles men did indeed have a cellular-immune dysfunction, one that would turn out to be caused by a virus known today as human immunodeficiency virus (HIV). The virus, researchers would later discover, had been secretly infecting victims for fifty years. During the 1980s it finally exploded, and since then it has infected sixty million people. It has killed nearly half of them.

HIV's death toll is all the more terrifying because it's actually not all that easy to catch. You can't get HIV if an infected person sneezes near you or shakes your hand. HIV has to be spread through certain bodily fluids, such as blood and semen. Unprotected sex can transmit the virus. Contaminated blood supplies can infect people through transfusions. Infected mothers can pass HIV to their unborn children. Many people who take heroin and other drugs have acquired HIV if they've shared needles with infected users.

Once HIV gets into a person's body, it boldly attacks the immune system itself. It grabs onto certain kinds of immune cells known as CD4 T cells and fuses their membranes like a pair of colliding soap bubbles. Like other retroviruses, it inserts its genetic material into the cell's own genome. Its genes and proteins manipulate then take over the cell, causing it to make new copies of HIV, which escape and can infect other cells.

At first, the population of HIV in a person's body explodes rapidly. Once the immune system recognizes infected cells it starts to kill them, driving the virus's population down. To the infected person, the battle feels like a mild flu. The immune system manages to exterminate most of the HIV, but a small fraction of the viruses manages to survive by lying low. The CD4 T cells in which they hid continued grow and divide. From time to time, an infected CD4 T cell wakes up and fires a blast of viruses that infect new cells. The immune system attacks these new waves, but over time it becomes exhausted and collapses.

It may take only a year for an immune system to fail, or more than twenty. But no matter how long it takes, the outcome is the same: people can no longer defend themselves against diseases that would never be able to harm a person with a healthy immune system. In the early 1980s, a wave of HIV-infected people began to come to hospitals with strange diseases like pneumocystis pneumonia.

Doctors discovered the effects of HIV before they discovered the virus, dubbing it acquired immunodeficiency syndrome, or AIDS. In 1983, two years after the first AIDS patients came to light, French scientists isolated HIV from a patient with AIDS for the first time. More research firmly established HIV as the cause of AIDS. Meanwhile, doctors were discovering more cases of AIDS, both in the United States and abroad. Other great scourges, such as malaria and tuberculosis, are ancient enemies, which had been killing people for thousands of years. Yet HIV went from utter obscurity in 1980 to a global scourge in a matter of a few years. Here was an epidemiological mystery.

To solve it, scientists began to sequence the genes of HIV they isolated from different patients. They examined HIV not just from the United States but from other countries around the world where it was beginning to spread as well. They drew evolutionary trees, with each strain of HIV a branch sprouting from a common ancestor. Researchers discovered that there was not one kind of HIV, but two. The vast majority of cases of HIV were caused by a strain that was dubbed HIV-1, and the rest were caused by a distinct form of the virus, called HIV-2. The two types of HIV could be distinguished in many ways, including the symptoms they caused: HIV-2 was much milder than HIV-1.

HIV, scientists found, belongs to a large group of slow-growing retroviruses, known as lentiviruses. Lentiviruses infect many mammals, including cats, horses, cows, and monkeys. In 1991, Preston Marx of New York University and his colleagues discovered that HIV-2 was closely related to lentiviruses that infect an African species of monkey called sooty mangabeys. They concluded that HIV-2 descended from a mangabey lentivirus. In West Africa, where HIV-2 is most common, some people keep the monkeys as pets; others eat them. Infected mangabeys may have introduced their lentivirus into humans with a bite.

It took scientists longer to pin down the origins of HIV-1, the strain that causes the vast majority of AIDS cases. That's because the closest relatives of HIV-1 lived in primates that are much harder to study: chimpanzees. Relatively few chimpanzees live in captivity, and trying to get blood samples from chimpanzees in the wild can be a staggeringly hard job. They're elusive, strong, and not fond of people with needles. Scientists had to develop new ways to test them for HIV, such as searching for the viruses in their feces. Slowly, scientists amassed a collection of HIV-1–like lentiviruses from chimpanzees. Comparing the viruses to each other, they discovered that some strains of HIV-1 are more closely related to certain chimpanzee viruses that they are to other HIV-1 strains. The branchings of the viral tree suggest that HIV-1 actually evolved from chimpanzee viruses several times.

But when did this transition happen? Some scientists tried to get an answer to that question by looking back at patients who had died mysteriously before the discovery of HIV. In 1988, for example, researchers discovered that a Norwegian sailor named Arvid Noe, who died in 1976, had HIV in his tissues. Reaching back further into HIV's history was nearly impossible, because many of its earliest victims lived in poor countries and died without any careful medical tests that would identify unusual diseases like pneumocystis pneumonia.

It turned out that the viruses replicating in living people offered some powerful clues to the origins of HIV. Through the 1990s, scientists at Los Alamos National Laboratory amassed a database of genetic sequences of HIV taken from thousands of patients. They could then use supercomputers to compare these viruses and figure out which mutations the viruses had acquired since they diverged from a common ancestor. By adding up these mutations, the researchers found that HIV gradually acquires mutations at a roughly regular rate. In other words, the mutations piled up like sand in an hourglass. By measuring how high the sand had piled up, they could estimate how much time had passed. They estimated that the common ancestor of HIV-1 existed in 1933.

That estimate has been confirmed by the remarkable discovery of HIV preserved in tissues stored away in hospitals in Kinshasa, the capital of the Democratic Republic of the Congo in central Africa. In 1998, David Ho and his colleagues at Rockefeller University reported that they had isolated HIV from a blood sample taken from a patient in Kinshasa in 1959. In 2008, Michael Worobey and his colleagues at the University of Arizona discovered HIV in a second tissue sample from another pathology collection in Kinshasa, dating back to 1960. These two samples allowed researchers to confirm that HIV emerged in the early 1900s.

The molecular clock created by the Los Alamos researchers was accurate enough to predict the age of the Kinshasa viruses based on their genetic sequence alone. But the two viruses also provide a surprising glimpse at the diversity of HIV in Kinshasa around 1960. Worobey and his colleagues found that the old viruses were not closely related to each other. Instead, they were each closely related to a different branch of HIV-1 found in patients today. Studying the distant kinship of these two viruses, Worobey and his colleagues concluded that all of the major branches of HIV-1 found in the world today already existed in 1960. What's more, they were all probably circulating around Kinshasa.

All this evidence is now pointing to how HIV-1 got its start. HIV-1-like viruses had been circulating among populations of chimpanzees throughout Africa. Hunters sometimes killed chimpanzees for meat, and from time to time they became infected by the viruses. But these hunters, living in relative isolation, were a dead end for the viruses. In the early 1900s the opportunities for these viruses changed as colonial settlements in central Africa began to expand to cities of ten thousand people or more. Commerce along the rivers allowed pathogens to reach the cities from remote forests. The chimpanzees carrying viruses most closely related to HIV-1 live today in the jungles of southeast Cameroon. It may be no coincidence that the rivers of that region flow south and eventually reach Kinshasa.

In the growing city of Kinshasa (then known as Leopoldville), HIV-1 could multiply. Instead of a few dead ends, it found a population that could sustain it and inside of which it could evolve into new forms better adapted to humans. By 1960, HIV-1 had bloomed into a wide genetic diversity, although it probably only infected a few thousand people.

Worobey and his colleagues have started to map the subsequent spread of HIV-1 out from Kinshasa to the rest of the world. The most common strain of HIV in the United States, for example, is known as HIV-1 subtype B. The oldest lineages of HIV-1 subtype B are found in Haiti, and Worobey estimates they branched off from African strains in the 1960s. That happens to be a time when many Haitians who had been working in the Congo returned to their homeland after the country became independent from Belgium. They may have unwittingly brought HIV back to the New World with them. Haitian immigrants or American tourists may have then brought HIV to the United States. The oldest lineages of HIV-1 subtype B in the United States, Worobey and his colleagues found, date back to about 1970. That's about four decades since the virus became established in humans, and about one decade before five men in Los Angeles became sick with a strange form of pneumonia.

By the time scientists recognized HIV in 1983, in other words, the virus had already begun to turn into a global catastrophe. As a result, HIV has had a huge head start on scientists who hope to halt its spread. It would not be until the early 1990s that some strategies began to show real promise for slowing the epidemic. Changing people's behaviors has proven effective. Uganda launched a major campaign against HIV that featured condom use and other public health measures. As a result, the country reduced its HIV rate from about 15 percent in the early 1990s to about 5 percent in 2001. Unfortunately funding for these programs began to ebb after a few years, and the infection rate in Uganda has begun to rise again.

Other researchers have investigated medications that can slow the rise of HIV in infected people, so that their immune systems can remain strong enough to block the onset of AIDS. Millions of people now take a cocktail of drugs that interfere with the ability of HIV to infect immune cells and use them to replicate. In affluent countries like the United States, these drug therapies have allowed some people to enjoy a relatively healthy life. But the cost of these drugs has meant that most people with HIV—living in the poorest countries—cannot afford a treatment that might give them extra years or even decades of life. That's beginning to change rapidly, as the United States and nongovernmental organizations are now starting to provide these drugs to the most afflicted countries and as treatment programs are starting to be scaled up dramatically.

Yet these drugs, even if they can prolong lives, are not the perfect cure. They have side effects that can become harmful after years of therapy, and they foster the evolution of resistant viruses, which then requires shifting patients to new drugs. In theory, the best solution to HIV would be a vaccine—either one that could prevent people from becoming infected with the virus or one that could stimulate the immune system of infected people to attack it effectively. Vaccines would be far less expensive than treating HIV infection with drug cocktails and could help slow down the transmission cycle. But the quest for an HIV vaccine has been a disappointing struggle so far. In 2008, for example, a highly anticipated trial of a vaccine developed by Merck had to be abandoned because the vaccine appeared to be making people more likely to acquire HIV, not less.

There's good reason to worry about any HIV vaccine, even one that shows promise in small trials. That's because HIV is evolving in overdrive. HIV belongs to a group of viruses—including influenza—that are very sloppy in their replication. They create many mutants in very little time. These mutants provide the raw material for natural selection to act on, producing viruses that are better and better adapted. Within a single host, natural selection can improve the ability of viruses to escape detection of the immune system.

In 2008, Philip Goulder, a medical researcher at Oxford, led an international team of scientists who found evidence for the ongoing evolution of HIV. They studied the immune systems of 2,800 people from all over the world, examining proteins known as human leukocyte antigens, which infected cells use to transport fragments of viruses to their surface. The fragments can then be recognized by immune cells, which destroys the infected cell. Different people carry different variations in the genes for human leukocyte antigens. Goulder and his colleagues found that most of the HIV in each country carried mutations to the most effective human leukocyte antigens in that country's population. Their findings tell us that HIV is rapidly adapting to the variations in human immune systems around the world. That is sobering news to those who are trying to build HIV vaccines. If a vaccine ever succeeds in boosting an effective immune response in people, HIV might well evolve a way to escape.

It's possible that vaccine developers could keep HIV from escaping by continually rolling out new vaccines that would stay one step ahead of the virus. Another intriguing possibility is to look back over its history. A team of American scientists compared a wide range of HIV-1 subtype B strains and reconstructed one of the proteins made by their common ancestor. They then used that ancestral protein to make a vaccine. The researchers found that monkeys injected with the vaccine were able to produce an immune response to a much wider range of HIV strains than more conventional vaccines. The future of fighting HIV, perhaps, may lie in its past.

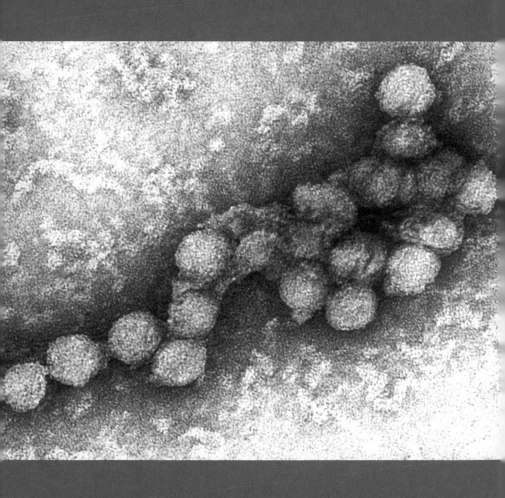

West Nile viruses in suspension

Becoming an American

West Nile Virus

In the summer of 1999, Tracey McNamara got worried. McNamara was the chief pathologist at the Bronx Zoo. When an animal at the zoo died, it was her job to figure out what killed it. She began to see dead crows on the ground near the zoo, and she wondered if they were being killed by some new virus spreading through the city. If the crows were dying, the zoo's animals might start to die too.

Over Labor Day weekend, her worst fears were realized: three flamingoes died suddenly. So did a pheasant, a bald eagle, and a cormorant. McNamara examined the dead birds and found they had all suffered bleeding in their brains. Their symptoms suggested that they had been killed by the same pathogen. But McNamara could not figure out what pathogen was responsible, so she sent tissue samples to government laboratories. The government scientists ran test after test for the various pathogens that might be responsible. For weeks, the tests kept coming up negative.

Meanwhile, doctors in Queens were seeing a worrying number of cases of encephalitis—an inflammation of the brain. The entire city of New York normally only sees nine cases a year, but in August 1999, doctors in Queens found eight cases in one weekend. As the summer waned, more cases came to light. Some patients suffered fevers so dire that they became paralyzed, and by September nine had died. Initial tests pointed to a viral disease called Saint Louis encephalitis, but later tests failed to match the results.

As doctors struggled to make sense of the human outbreak, McNamara was finally getting the answer to her own mystery. The National Veterinary Services Laboratory in Iowa managed to grow viruses from the bird tissue samples she had sent them from the zoo. They bore a resemblance to the Saint Louis encephalitis virus. McNamara wondered now if both humans and birds were succumbing to the same pathogen. She convinced the Centers for Disease Control and Prevention to analyze the genetic material in the viruses. On September 22, the CDC researchers were stunned to find that the birds were not killed by Saint Louis encephalitics. Instead, the culprit was a pathogen called West Nile virus, which infects birds as well as peolple in parts of Aisa, Europe, and Africa. No one had imagined that the Bronx Zoo birds were dying of West Nile virus, because it had never been seen in a bird in the Western Hemisphere before.

Public health workers puzzling over the human cases of encephalitics decided it was time to broaden their search as well. Two teams—one at the CDC and another led by Ian Lipkin, who was then at the University of Californa, Irvine—isolated the genetic material from the human viruses. It was the same virus that was killing birds: West Nile. And once again, it took researchers by surprise. No human in North or South America had ever suffered from it before.

The United States is home to many viruses that make people sick. Some are old and some are new. When the first humans made their way into the Western Hemisphere some fifteen thousand years ago, they brought a number of viruses with them. Human papillomavirus, for example, retains traces of its ancient emigration. The strains of the virus found in Native Americans are more closely related to each other than they are to HPV strains in other parts of the world. Their closest relative outside of the New World are strains of HPV found in Asia, just as Native Americans are most closely related to Asians.

Columbus's discovery of the New World triggered a second wave of new viruses. Europeans brought viruses causing diseases such as influenza and smallpox that wiped out most Native Americans. In later centuries, still more viruses arrived. HIV came to the United States in the 1970s, and at the end of the twentieth century, West Nile virus became one of America's newest immigrants.

It had only been six decades since West Nile virus was discovered anywhere on the planet. In 1937, a woman in the West Nile district of Uganda came to a hospital with a mysterious fever, and her doctors isolated a new virus from her blood. Over the next few decades, scientists found the same virus in many patients in the Near East, Asia, and Australia. But they also discovered that West Nile virus did not depend on humans for its survival. Researchers detected the virus in many species in birds, where it could multiply to far higher numbers.

At first it was not clear how the virus could move from human to human, from bird to bird, or from bird to human. That mystery was solved when scientists found the virus in a very different kind of animal: mosquitoes. When a virus-bearing mosquito bites a bird, it sticks its syringe-like mouth into the animal's skin. As the mosquito drinks, it squirts saliva into the wound. Along with the saliva comes the West Nile virus.

The virus first invades cells in the bird's skin, including immune system cells that are supposed to defend animals from diseases. Virus-laden immune cells crawl into the lymph nodes, where they release their passengers, leading to the infection of more immune cells. From the lymph nodes, infected immune cells spread into the bloodstream and organs such as the spleen and kidneys. It takes just a few days for the viruses in a mosquito bite to multiply into billions inside a bird. Despite their huge numbers, West Nile viruses cannot escape a bird on their own. They need a vector. An mosquito must bite the infected bird, drawing up some of its virus-laden blood. Once in the mosquito, the viruses invade the cells of its midgut. From there they can be carried to the insect's salivary glands, where the viruses are ready to be injected into a new bird.

Vector-borne viruses like West Nile virus require a special versatility to complete their life cycle. Mosquitoes and birds are profoundly different kinds of hosts, with different body temperatures, different immune systems, and different anatomies. West Nile virus has to be able to thrive in both environments to complete its life cycle. Vector-borne viruses also pose special challenges to doctors and public health workers who want to stop their spread. They don't require people to be in close contact to spread from host to host. Mosquitoes, in effect, give the viruses wings.

Studies on the genes of West Nile virus suggest that it first evolved in Africa. As birds migrated from Africa to other continents in the Old World, they spread the virus to new bird species. Along the way, West Nile virus infected humans. In Eastern Europe, epidemics broke out, producing some cases of encephalitis. In a 1996 epidemic in Romania, ninety thousand people came down with West Nile, leading to seventeen deaths. These new epidemics, first in Europe and later in the West, may have been the result of the virus infecting people who populations had not experienced it before. In Africa, by contrast, people may be immunized against West Nile virus after being infected while they're young.

It is striking that the New World has been spared West Nile virus for so long. The flow of people across the Atlantic and Pacific was not enough to carry the virus to the Americas. Scientists cannot say exactly how West Nile virus finally landed in New York in 1999, but they have a few clues. The New World strain of West Nile virus is most closely related to viruses that caused an outbreak in birds in Israel in 1998. It's possible that pet smugglers brought infected birds from the Near East to New York.

On its own, a single infected bird could not have triggered a nationwide epidemic. The viruses needed a new vector to spread. It just so happens that West Nile viruses can survive inside 62 species of mosquitoes that live in the United States. The birds of America turned out to be good hosts as well. All told, 150 American bird species have been found to carry West Nile virus. A few species, such as robins, blue jays, and house finches, turned out to be particularly good incubators.

Moving from bird to mosquito to bird, West Nile virus spread across the entire United States in just 4 years. And along the way, people became ill with West Nile virus as well. About 85 percent of infections in the United States cause no symptoms. The other 15 percent of infected people develop fevers, rashes, and headaches, and 38 percent of them have to go to a hospital, where they stay for about 5 days on average. About 1 in 150 infected people end up developing encephalitis. Between 1999 and 2008, U.S. doctors recorded 28,961 cases of West Nile virus. Of those victims, 1,131 died.

Once West Nile virus arrived in the United States, it settled into a regular cycle, a cycle set by the natural history of birds and mosquitoes. In the spring, robins and other birds produce new generations of chicks that are helpless targets for virus-carrying mosquitoes. By the summer, many birds are positively brimming with West Nile virus, raising the fraction of mosquitoes that carry it. It's at that time of year that most human cases of West Nile virus emerge. When the temperature falls, mosquitoes die, and the viruses can no longer spread. It's not clear how the virus survives North American winters. It's possible that they survive in low levels among mosquitoes in the south, where the winters aren't so harsh. It's also possible that mosquitoes infect own their eggs with West Nile virus. When infected eggs hatch the next spring, the new generation is ready to start infecting birds all over again.

West Nile virus has fit so successfully into the ecology of the United States that it's probably going to be impossible to eradicate. Unfortunately, doctors have no vaccine to prevent West Nile virus and no drugs to treat an infection. If you get sick, you can only hope that you are among the majority who suffer a fever and then recover. And in the future, West Nile virus may become even more entrenched in its new home. Jonathan Soverow of Beth Israel Deaconess Medical Center and his colleagues examined sixteen thousand cases of West Nile virus that occurred between

2001 and 2005, noting the weather at the time of each outbreak. They found that epidemics tended to occur when there was heavy rainfall, high humidity, and warm temperatures. Warm, rainy, muggy weather makes mosquitoes reproduce faster and makes their breeding season longer. It also speeds up the growth of the viruses inside the mosquitoes.

Unfortunately, we can expect more of that sort of weather in the future. Carbon dioxide and other heat-trapping gases are raising the average temperature in the United States, and climate scientists project that the temperature will continue to rise much higher in decades to come. Now that West Nile virus has made a new home here, we're making that home more comfortable.

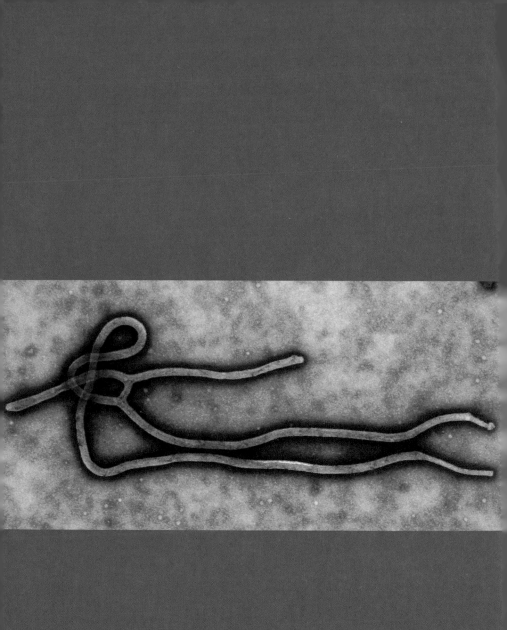

Ebola virus in suspension

Predicting the Next Plague

Severe Acute Respiratory Syndrome
and Ebola

A hunter emerges from a tropical forest, a shotgun in one hand, the carcass of a monkey in the other. He walks into a village in the southeast corner of Cameroon. It's a scene that replays itself every day in villages, not just in Africa, but around the world. Hunters kill wild animals and bring them home to feed their families or to sell for cash. But on this day, the scene ends with a twist. The hunter hands over the monkey to his wife to butcher. As she cuts up the monkey, she stops to hold a dismembered leg over a piece of paper marked with five circles. Drops of blood fill one circle after another. The hunter's wife then slips the sheet of paper in a Ziploc bag and hands it to a team of scientists who have paid her a visit. The scientist, who belongs to an organization called the Global Viral Forecasting Initiative, will analyze the blood-soaked paper for viruses infecting the monkey.

The Global Viral Forecasting Initiative is trying to change the way we fight viruses. Someday, somewhere, a virus we don't know about is going to emerge as a major new threat to human health. We've seen it happen many times before, and so we know it will happen again. GVFI scientists think we'll do a better job fighting that new virus if we can learn something about it in advance. To eliminate the advantage of surprise, GVFI scientists are looking for these viruses before they jump into humans. The best place to look for them is in animals, such as the monkeys that Cameroonian hunters kill for food.

The threat of new viruses has inspired a string of cheesy Hollywood movies over the years. In *The Andromeda Strain*, which came out in 1971, a satellite falls to Earth with an extraterrestrial virus that threatens to wipe out humanity. In the 1995 movie *Outbreak*, a monkey imported from Africa spreads a deadly virus through a California town, which the Army wants to bomb to prevent it from spreading across the country. And in *28 Days Later*, released in 2002, a virus sweeps through London, turning its victims into homicidal maniacs.

The reality of new viruses is nothing like these fantasies. In its own way, it's far more frightening. Over the course of human history, many viruses have made the evolutionary leap from animal hosts to our own species. And just over the past century, dozens of viruses have made this transition, giving rise to new diseases. Scientists have found that these new viruses have generally taken the same route into our species. It's likely that they will take the same path in the future.

Many human viruses evolved from ancestral pathogens that were well adapted to living in other species. For example, HIV evolved from a virus found in chimpanzees known as SIVcpz. For centuries, the virus moved from chimpanzee to chimpanzee, infecting immune cells and slowly eroding their defenses. In the early 1900s, some of the viruses moved from chimpanzees to humans, evolving into HIV. The most HIV-like strains of SIVcpz are carried by chimpanzees that live in the forests in Cameroon. It was there that the virus likely made the transition. Both SIVcpz and HIV are spread through blood-to-blood contact. SIVcpz probably first infected the hunters who killed chimpanzees for meat. The virus-laden blood in the butchered apes made contact with cuts on the hunters, delivering SIVcpz into new hosts.

When animal viruses first make contact with humans, they only use them as what scientists called "spillover hosts." Adapted to growing in other animals, the viruses can only grow slowly in humans and typically fail to spread from one human to another. When SIVcpz started infecting hunters, it probably still depended on chimpanzees to replenish its numbers. But the viruses were also mutating rapidly, and mutant SIVcpz eventually evolved the ability to survive in humans and spread from one human to the next.

Initially, new human viruses only cause local outbreaks, because they still can't move between people very well. After each human epidemic sputters to an end, the virus still thrives in its animal host. But as the virus spends more time in humans, natural selection favors mutations that adapt them to their new host. The epidemics in humans get bigger and last longer. HIV, for example, thrived as African colonies grew and networks of roads linked forest villages to large cities where the virus could circulate among many people. As HIV became better adapted to infecting humans, it lost its ability to attack chimpanzees.

No one knew about the transformation of HIV while it was happening. Only in the early 1980s, sixty years or so after the virus had crossed into our species, did scientists finally isolate the virus and realize it was causing AIDS. By then, HIV was well established in our species and started to become one of the worst diseases in human history. We can only speculate about how much easier it would have been to fight the disease back when it was infecting just a few hundred villagers in Cameroon.

In recent years, scientists have been able to identify new human diseases far faster. In November 2002, for example, a Chinese farmer came to a hospital suffering from a high fever and died soon afterward. Other people from the same region of China began to develop the disease as well, but it didn't reach the world's attention until an American businessman flying back from China developed a fever on a flight to Singapore. The flight stopped in Hanoi, where the businessman died. Soon, people were falling ill in countries around the world, although most of the cases turned up in China and Hong Kong. About 10 percent of people who became sick died in a matter of days. The disease was not one that any doctor had identified before—not the flu, not pneumonia, nor any other known disease. It was dubbed severe acute respiratory syndrome, or SARS.

Scientists began searching samples from SARS victims for a cause of the diseases. Malik Peiris of the University of Hong Kong led the team of researchers who found it. In a study of fifty patients with SARS, they discovered a virus growing in two of them. The virus belonged to a group of species called coronaviruses, which can cause colds and the stomach flu. Peiris and his colleagues sequenced the genetic material in the new virus and then searched for matching genes in the other patients. They found a match in forty-five of them.

Based on their experience with viruses such as HIV, scientists suspected that the SARS virus had evolved from a virus that infects animals. They began to analyze viruses in animals with which people in China have regular contact. As they discovered new viruses, they added their branches to the SARS evolutionary tree. In a matter of months, scientists had reconstructed the history of SARS.

The virus started in Chinese bats. A lineage of the viruses then began to spill over into a catlike mammal called a civet. Civets are a common sight in Chinese animal markets, and it's likely that humans became spillover hosts as well. The virus then evolved the ability to leap from human to human. SARS was a very young virus when scientists discovered it, and the speed at which it was discovered helped make it a relatively small outbreak. Scientists were able to identify and quarantine people with the disease, and they banned the sale of civets in markets. Although the SARS virus managed to spread across much of the world, it only caused about eight thousand cases and nine hundred deaths before it disappeared.

We can expect more viruses to sweep into our species, and they will probably come at an accelerating pace. Animals in remote parts of the world have harbored exotic viruses for millions of years, and for all that time humans have had little contact with them. Now humans are moving deep into these remote territories to harvest timber, dig mines, and establish new farms. And in the process, they've come into contact with new viruses. Nipah virus, for example, causes dangerous inflammation of the brain in its victims in Southeast Asia. It's a virus that normally lives in bats, which once lived far from humans in jungles. Now the bats—and the viruses—have no jungles to live in.

There's no reason to think that one of these new viruses will wipe out the human race. That may be the impression that movies like *The Andromeda Strain* give, but the biology of real viruses suggests otherwise. Ebola, for example, is a horrific virus that can cause people to bleed from all their orifices, including their eyes. It can sweep from victim to victim, killing almost all its hosts along the way. And yet a typical Ebola outbreak only kills a few dozen people before coming to a halt. The virus is just too good at making people sick, and so it kills its victims faster than it can find new ones. Once an Ebola outbreak ends, the virus vanishes for years.

Ebola-like viruses may be frightening, but they probably pose less of a danger to our species than viruses with a lower death rate that can spread to more hosts. The 1918 outbreak of influenza killed only a tiny fraction of its victims. But because it infected one in three people on Earth, that tiny fraction added up to an estimated fifty million people. HIV crept slowly and surreptitiously around the planet before it was first detected. Instead of causing the terrifying symptoms of Ebola, HIV quietly breaks down the immune system over the course of many years.

We don't know which virus will create the next great epidemic, in part because we don't know the world of viruses very well. GVFI scientists have discovered a number of new viruses in African monkeys. Their tests on hunters have revealed those viruses in humans as well. Fortunately, these new viruses cannot yet spread from person to person. But that doesn't mean that we can simply ignore them. Just the opposite: these are the viruses we need to block before they get a chance to make the great leap into our species.

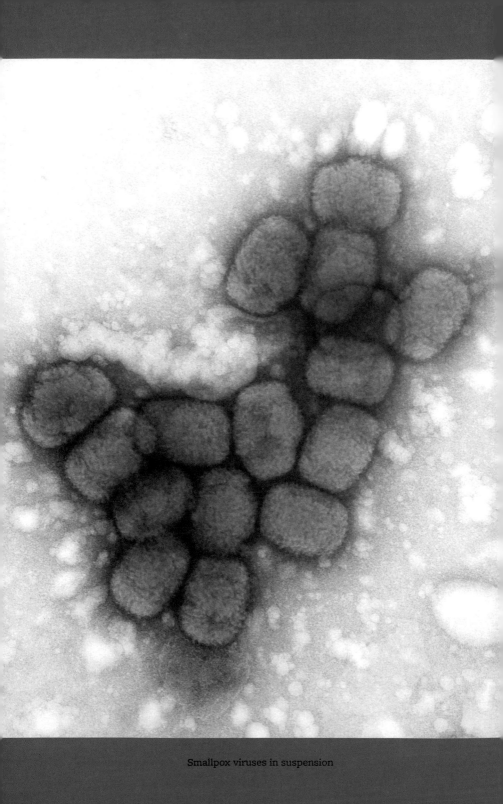

Smallpox viruses in suspension

The Long Goodbye

Smallpox

We humans are good at creating new viruses by accident—whether it's a new flu virus concocted on a pig farm, or HIV evolving from the viruses of butchered chimpanzees. What we're not so good at is getting rid of viruses. Despite all the vaccines, antiviral drugs, and public health strategies at our disposal, viruses still manage to escape annihilation. The best we can typically manage is to reduce the harm viruses cause. HIV infections, for example, have declined in the United States, but fifty thousand Americans still acquire the virus every year. Vaccination programs have eliminated some viruses from some countries, but the viruses can still thrive in other parts of the world. In fact, modern medicine has only managed to completely eradicate a single species of human virus from nature. The distinction goes to the virus that causes smallpox.

But what a virus to wipe out. Over the past three thousand years, smallpox may have killed more people than any other disease on Earth. Ancient physicians were well aware of smallpox, because its symptoms were so clear and distinct. A victim became infected when the virus slipped into the airway. After a week or so, the infection brought chills, a blazing fever, and agonizing aches. The fever ebbed after a few days, but the virus was far from done. Red spots developed inside the mouth, then the face, and then over the rest of the body. The spots filled with pus and caused stabbing pain. About a third of people who got smallpox eventually died. In the survivors, scabs covered over the pustules, which left behind deep, permanent scars.

Some thirty-five hundred years ago, smallpox left its first recorded trace on humanity: three mummies from ancient Egypt, studded with pustules. Many of the oldest centers of civilization in the Old World, from China to India to ancient Greece, felt the wrath of the virus. In 430 BC, an epidemic of smallpox swept through Athens, killing a quarter of the Athenian army and a large percentage of the city's population. In the Middle Ages, crusaders returning from the Middle East brought smallpox to Europe. Each time the virus arrived in a new defenseless population, the effects were devastating. In 1241 smallpox first came to Iceland, where it promptly killed twenty thousand of the island's seventy thousand inhabitants. Smallpox became well established in the Old World as cities grew, providing a high density of potential hosts. Between 1400 and 1800, smallpox killed an estimated five hundred million people every century in Europe alone. Its victims included sovereigns such as Czar Peter II of Russia, Queen Mary II of England, and Emperor Joseph I of Austria.

It was not until Columbus's arrival in the New World that Native Americans got their first exposure to the virus. The Europeans unwittingly brought a biological weapon with them that gave the invaders a brutal advantage over their opponents. With no immunity whatsoever to smallpox, Native Americans died in droves when they were exposed to the virus. In Central America, over 90 percent of the native population is believed to have died of smallpox in the decades following the arrival of the Spanish conquistadores in the early 1500s.

The first effective way to prevent the spread of smallpox probably arose in China around AD 900. A physician would rub a scab from a smallpox victim into a scratch in the skin of a healthy person. (Sometimes they administered it as an inhaled powder instead.) Variolation, as this process came to be called, typically caused just a single pustule to form on the inoculated arm. Once the pustule scabbed over, a variolated person became immune to smallpox.

At least, that was the idea. Fairly often, variolation would trigger more pustules, and in 2 percent of cases, people died. Still, a 2 percent risk was more attractive than the 30 percent risk of dying from a full-blown case of smallpox. Variolation spread across Asia, moving west along trade routes until the practice came to Constantinople in the 1600s. As news of its success traveled into Europe, physicians there began to practice variolation as well. The practice triggered religious objections that only God should decide who survived the dreaded smallpox. To counteract these suspicions, doctors organized public experiments. Zabdiel Boylston, a Boston doctor, publicly variolated hundreds of people in 1721 during a smallpox epidemic; those who had been variolated survived the epidemic in greater numbers than those who had not been part of the trial.

No one at the time knew why variolation worked, because nobody knew what viruses were or how our immune systems fought them. The treatment of smallpox moved forward mainly by trial and error. In the late 1700s, the British physician Edward Jenner invented a safer smallpox vaccine based on stories he heard about how milkmaids never got smallpox. Cows can get infected with cowpox, a close relative of smallpox, and so Jenner wondered if it provided some protection. He took pus from the hand of a milkmaid named Sarah Nelmes and inoculated it into the arm of a boy. The boy developed a few small pustules, but otherwise he suffered no symptoms. Six weeks later, Jenner variolated the boy—in other words, he exposed the boy to smallpox, rather than cowpox. The boy developed no pustules at all. Jenner published a pamphlet in 1798 documenting this new, safer way to prevent smallpox. He dubbed it "vaccination," after the Latin name of cowpox, *Variolae vaccinae*. Within three years, over one hundred thousand people in England had gotten vaccinated against smallpox, and vaccinations spread around the world. In later years, other scientists borrowed Jenner's techniques and invented vaccines for other viruses. From rumors about milkmaids came a medical revolution.

As vaccines grew popular, doctors struggled to keep up with the demand. At first they would pick off the scabs that formed on vaccinated arms, and use them to vaccinate others in turn. But since cowpox occurred naturally only in Europe, people in other parts of the world could not simply acquire the virus themselves. In 1803, King Carlos of Spain came up with a radical solution: a vaccine expedition to the Americas and Asia. Twenty orphans boarded a ship in Spain. One of the orphans had been vaccinated before the ship set sail. After eight days, the orphan developed pustules, and then scabs. Those scabs were used to vaccinate another orphan, and so on through a chain of vaccination. As the ship stopped in port after port, the expedition delivered scabs to vaccinate the local population.

Physicians struggled throughout the 1800s to find a better way to deliver smallpox vaccines. Some turned calves into vaccine factories, infecting them repeatedly with cowpox. Some experimented with preserving the scabs in fluids like glycerol. It wasn't until scientists finally worked out the nature of smallpox and cowpox—the fact that they were viruses—that it became possible to develop a vaccine that could be made on an industrial scale and shipped around the world.

Once vaccines became common, smallpox began to lose its fierce grip on humanity. Through the early 1900s, one country after another recorded their last case of smallpox. By 1959, smallpox had retreated from Europe, the Soviet Union, and North America. It remained a scourge of tropical countries with poor medical systems in place. But it was beaten so far back that some public health workers began to contemplate an audacious goal: eliminating smallpox from the planet altogether.

The advocates of smallpox eradication built their case on the biology of the virus. Smallpox only infects humans, not animals. If it could be systematically eliminated from every human population, there would be no need to worry that it was lurking in pigs or ducks, waiting to reinfect us. What's more, smallpox is an obvious disease. Unlike a virus like HIV, which can take years to make itself known, smallpox declares its gruesome presence in a matter of days. Public health workers would be able to identify outbreaks and track them with great precision.

Yet the idea of eradicating smallpox met with intense skepticism. If everything went exactly according to plan, an eradication project would require years of labor by thousands of trained workers, spread across much of the world, toiling in many remote, dangerous place. Public health workers had already tried to eradicate other diseases, like malaria, and failed.

The skeptics lost the debate, however, and in 1965, the World Health Organization launched the Intensified Smallpox Eradication Programme. The eradication effort was different in many ways from previous campaigns. It relied on a new prong-shaped needle that could deliver smallpox vaccine far more efficiently than regular syringes. As a result, vaccine supplies could be stretched much further than before. Public health workers also designed smart new strategies for administering vaccines. Trying to vaccinate entire countries was beyond the reach of the eradication project. Instead, public health workers identified outbreaks and took quick action to snuff them out. They quarantined victims and then vaccinated people in the surrounding villages and towns. The smallpox would spread like a forest fire, but soon it would hit the firebreak of vaccination and die out.

Outbreak by outbreak, the virus was beaten back, until the last case was recorded in Ethiopia in 1977. The world was now free of smallpox.

While the eradication campaign was a huge success, the smallpox virus had not disappeared completely. Scientists had established stocks of the virus in their laboratories to study. The WHO had all the stocks gathered up and deposited in two approved laboratories, one in the Siberian city of Novosibirsk in the Soviet Union, and one at the U.S. Centers for Disease Control and Prevention in Atlanta, Georgia. Smallpox experts could still study stocks from the two labs, but only under tight regulations. Most experts assumed that before long those last two collections of smallpox would be destroyed as well, and then the virus would become truly extinct.

It turns out, however, that there might actually be more smallpox virus in the world. In the 1990s, Soviet defectors revealed that their government had actually set up labs to produce a weaponized smallpox virus that could be loaded onto missiles and launched at enemy targets. After the fall of the Soviet government, the labs were abandoned. No one knows what ultimately happened to all the stocks of smallpox virus. We are left with the terrifying possibility that ex-Soviet virologists sold smallpox stocks to other governments or even terrorist organizations.

When these revelations emerged, some scientists and government officials decided the research stocks had to be preserved. Scientists could study them to help prepare for biological warfare. There remains much scientists don't yet understand about smallpox. In recent years, scientists have started to decipher the strategies smallpox uses to fight the immune system. They have discovered an arsenal of weapons the virus deploys. Smallpox proteins can jam the signals the immune cells pass to each other to mobilize an attack, for example. Scientists have yet to figure out why smallpox is so deadly. Some researchers argue that the virus causes the immune system to attack a victim's own body, rather than the virus. But that's just a hypothesis still to be tested. Solving mysteries like these could conceivably lead to better vaccines, and even to antiviral drugs that might be effective against smallpox infections or other dangerous viruses that are equally deadly to humans.

In 2010, the WHO reopened the debate over whether to finally destroy the two remaining officially declared stocks smallpox in Russia and the United States. But now the debate has taken a twist that previous generations of smallpox fighters could never have dreamed of. Today scientists know the full genetic sequence of the smallpox virus. And they have the technology necessary to synthesize the smallpox genome from scratch. Synthesizing viruses is not the stuff of science fiction; scientists have already manufactured the genetic material of other viruses, like polio and the deadly 1918 influenza, and have used it to generate full-blown viruses.

There's no evidence that anyone has tried to resurrect smallpox in the same way, but, then again, there's no evidence that it would be impossible to do so. After thirty-five hundred years of suffering and puzzling over smallpox, we have finally figured it out. And yet, by understanding smallpox, we have ensured that it can never be utterly eradicated as a threat to humans. Our knowledge gives the virus its own kind of immortality.

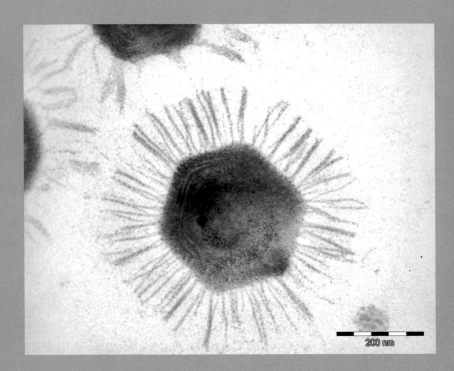

Mimivirus: the largest known virus

Epilogue

The Alien in the Watercooler

Mimivirus

Wherever there is water on Earth, there is life. The water may be a pond in a tropical forest, a pool in the Cave of Crystals, or a cooling tower sitting on the roof of a hospital.

In 1992, a microbiologist named Timothy Rowbotham scooped up some water from a hospital cooling tower in the English city of Bradford. He put it under a microscope and saw a welter of life. He saw amoebae and other single-celled protozoans, about the size of human cells. He saw bacteria, about a hundred times smaller. Rowbotham was searching for the cause of an outbreak of pneumonia that had been raging through Bradford. In the ranks of the microbes he found in the cooling tower water, he thought he found a promising candidate: a sphere of bacterial size, sitting inside an amoeba. Rowbotham believed he had found a new bacterium, and dubbed it *Bradfordcoccus*.

Rowbotham spent years trying to make sense of *Bradfordcoccus*, to see if it was the culprit in the pneumonia outbreak. He tried to isolate its genes by searching for stretches of DNA found in all bacteria, but he couldn't find any. Budget cuts forced Rowbotham to close his lab down in 1998, and so he arranged for French colleagues to store his samples. For five years, *Bradfordcoccus* languished in obscurity, until Bernard La Scola of Mediterranean University, decided to take another look at it. As soon as he put Rowbotham's samples under a microscope, he realized something was not right.

Bradfordcoccus did not have the smooth surface of spherical bacteria. Instead, it was more like a soccer ball, made up of many interlocking plates. And radiating out from its geometric shell La Scola saw hairlike threads of protein. The only things in nature that have these kinds of shells and threads were viruses. But La Scola knew, like all microbiologists at the time knew, that something the size of *Bradfordcoccus* was a hundred times too big to be a virus.

Yet a virus is exactly what *Bradfordcoccus* turned out to be. La Scola and his colleagues discovered that it reproduced by invading amoebae and forcing them to build new copies of itself. Only viruses reproduce this way. La Scola and his colleagues gave *Bradfordcoccus* a new name to reflect its viral nature. They called it a mimivirus, in honor of the virus's ability to mimic bacteria.

The French scientists then set out to analyze the genes of the mimivirus. Rowbotham had tried—and failed—to match its genes to those of bacteria. The French scientists had better luck. The mimivirus had virus genes—and a lot of them. Before the discovery of mimiviruses, scientists were used to finding only a few genes in a virus. But mimiviruses have 1,262 genes. It was as if someone took the genomes of the flu, the cold, smallpox, and a hundred other viruses and stuffed them all into one protein shell. The mimivirus even had more genes than some species of bacteria. In both its size and its genes, mimivirus had broken cardinal rules for being a virus.

Once La Scola and his colleagues knew what mimivirus genes looked like, they began to search for them in other habitats. They found the giant viruses in the lungs of hospital patients suffering from pneumonia. It's not clear yet if mimiviruses actually cause pneumonia, as Rowbotham had originally suspected, or if they just colonize people who are already sick. Scientists have also found mimiviruses and related giant viruses far from hospitals. They are actually common in the world's oceans, where they infect algae and perhaps even corals and sponges. Until now, scientists have realized, these giant viruses have been hiding in plain sight.

Newly discovered viruses like the mimivirus are forcing scientists to rethink what it means to be a virus in the first place. Their old rules, once so ironclad, are buckling. And as scientists debate what it means to be a virus, they are debating an even bigger question: what it means to be alive.

Scientists have long seen a huge gulf dividing viruses from "true" living things—bacteria, protozoans, plants, animals, and fungi. Many pointed to the tiny number of genes in viruses, arguing that there was no way for them to gain more because of their peculiar way of reproducing. Because viruses hijack cells to make new viruses, they are sloppy about copying their genes. They don't carry their own repair enzymes that can fix errors, for example. As a result, they are much more vulnerable to lethal mutations. If a virus accumulated thousands of genes, its high mutation rate would wipe it out.

The sizes of virus genomes offered some good reason to believe this was actually true. Viruses carry genes encoded either in DNA, or its single-stranded version, RNA. For a number of reasons, RNA is an inherently more error-prone molecule to copy. And it turns out that RNA viruses, like influenza and HIV, have smaller genomes than DNA viruses.

Forced to carry tiny genomes, viruses could not make room for genes that did anything beyond make new viruses and help those viruses escape destruction. They could carry genes to let them eat, for example. They could not turn raw ingredients into new genes and proteins on their own. They could not grow. They could not expel waste. They could not defend against hot and cold. They could not reproduce by splitting in two. All those *nots* added up to one great, devastating *Not*. Viruses were not alive.

To be alive, many scientists argued, required having a true cell. "An organism is constituted of cells," the microbiologist Andre Lwoff declared in a lecture he gave when he accepted the Nobel Prize in 1967. Lacking cells, viruses were considered as little more than cast-off genetic material that happened to have the right chemistry to get replicated inside cells that were truly alive. Scientists could purify viruses down to crystals, the same way they could crystallize salt or pure DNA. No one could ever crystallize a maple tree. In 2000, the International Committee on Taxonomy of Viruses declared that "viruses are not living organisms."

In the decade following that declaration, a number of scientists rejected it outright. The old rules no longer work well in the face of new viruses. Mimiviruses, for example, went overlooked for so long in part because they were a hundred times bigger than viruses are supposed to be. They are also loaded with far too many genes to fit old-fashioned notions of a virus. Scientists don't know what mimiviruses do with all of their genes, but some suspect that they do some rather lifelike things with them. Some of their proteins, for instance, look a lot like the proteins our own cells use to assemble new genes and proteins. When mimiviruses invade amoebae, they don't dissolve into a cloud of molecules. Instead, they set up a massive, intricate structure called a viral factory. The virus factory takes in raw ingredients through one portal, and then spits out new DNA and proteins through two others. The viral factory looks and acts remarkably like a cell. It's so much like a cell, in fact, that La Scola and his colleagues discovered in 2008 that it can be infected by a virus of its own. It was the first time anyone had found a virus of a virus. It was yet another thing that ought not to exist.

Drawing dividing lines through nature can be scientifically useful, but when it comes to understanding life itself, those lines can end up being artificial barriers. Rather than trying to figure out how viruses are not like other living things, it may be more useful to think about how viruses and other organisms form a continuum. We humans are an inextricable blend of mammal and virus. Remove our virus-derived genes, and we would be unable to reproduce. We would probably also quickly fall victim to infections from other viruses. Some of the oxygen we breathe is produced through a mingling of viruses and bacteria in the oceans. That mixture is not a fixed combination, but an ever-changing flux. The oceans are a living matrix of genes, shuttling among hosts and viruses.

Drawing a bright line between life and nonlife can also make it harder to understand how life began in the first place. Scientists are still trying to work out the origin of life, but one thing is clear: it did not start suddenly with the flick of a great cosmic power switch. It's likely that life emerged gradually, as raw ingredients like sugar and phosphate combined in increasingly complex reactions on the early Earth. It's possible, for example, that single-stranded molecules of RNA gradually grew and acquired the ability to make copies of themselves. Trying to find a moment in time when such RNA-life abruptly became "alive" just distracts us from the gradual transition to life as we know it.

Banning viruses from the Life Club also deprives us of some of the most important clues to how life began. One of the great discoveries about viruses has been the tremendous diversity in their genes. Every time scientists find new viruses, most of their genes bear little resemblance to any gene ever found before. The genes of viruses are not a meager collection of DNA cast off in recent years from true living things. Many scientists now argue that viruses contain a genetic archive that's been circulating the planet for billions of years. When they try to trace the common ancestry of virus genes, they often work their way back to a time before the common ancestor of all cell-based life. Viruses may have first evolved before the first true cells even existed. At the time, life may have consisted of little more than brief coalitions of genes, which sometimes thrived and sometimes were undermined by genes that acted like parasites. Patrick Forterre, a French virologist, has even proposed that in the RNA world, viruses invented the double-stranded DNA molecule as a way to protect their genes from attack. Eventually their hosts took over their DNA, which then took over the world. Life as we know it, in other words, may have needed viruses to get its start.

At long last, we may be returning to the original two-sided sense of the word *virus*, which originally signified either a life-giving substance or a deadly venom. Viruses are indeed exquisitely deadly, but they have provided the world with some of its most important innovations. Creation and destruction join together once more.

Acknowledgments

A *Planet of Viruses* was funded by the National Center for Research Resources at the National Institutes of Health through the Science Education Partnership Award (SEPA), grant no. R25 RR024267 (2007–2012), Judy Diamond, Moira Rankin and Charles Wood, principal investigators. Its content is solely the responsibility of the author and does not necessarily represent the official views of the NCRR or the NIH. I thank the many people who advised this project: Anisa Angeletti, Peter Angeletti, Aaron Brault, Ruben Donis, Ann Downer-Hazell, David Dunigan, Angie Fox, Laurie Garrett, Benjamin David Jee, Ian Lipkin, Ian Mackay, Grant McFadden, Nathan Meier, Abbie Smith, Gavin Smith, Philip W. Smith, Amy Spiegel, David Uttal, James L. Van Etten, Kristin Watkins, Willie Wilson, and Nathan Wolfe. I am particularly grateful to my SEPA program officer, L. Tony Beck, and to my editor at the University of Chicago Press, Christie Henry, for making this book possible.

Selected References

INTRODUCTION

Bos, L. 1999. Beijerinck' s work on tobacco mosaic virus: Historical context and legacy. *Philosophical Transactions of the Royal Society B: Biological Sciences* 354:675.

Flint, S. J. 2009. *Principles of virology*. 3rd ed. Washington DC: ASM Press.

Willner D., M. Furlan, M. Haynes, et al. 2009. Metagenomic analysis of respiratory tract DNA viral communities in cystic fibrosis and non-cystic fibrosis individuals. *PLoS ONE* 4 (10): e7370.

THE UNCOMMON COLD

Arden, K. E., and I. M. Mackay. 2009. Human rhinoviruses: Coming in from the cold. *Genome Medicine* 1:44.

Briese, T., N. Renwick, M. Venter, et al. 2008. Global distribution of novel rhinovirus genotype. *Emerging Infectious Diseases* 14:944.

Palmenberg, A. C., D. Spiro, R. Kuzmickas, et al. 2009. Sequencing and analyses of all known human rhinovirus genomes reveal structure and evolution. *Science* 324:55–59.

Simasek, M., and D. A. Blandino. 2007. Treatment of the common cold. *American Family Physician* 75:515–20.

LOOKING DOWN FROM THE STARS

Barry, J. M. 2004. *The great influenza: The epic story of the deadliest plague in history*. New York: Viking.

Dugan, V. G., R. Chen, D. J. Spiro, et al. 2008. The evolutionary genetics and emergence of avian influenza viruses in wild birds. *PLoS Pathogens* 4 (5): e1000076.

Rambaut, A., O. G. Pybus, M. I. Nelson, C. Viboud, J. K. Taubenberger, and E. C. Holmes. 2008. The genomic and epidemiological dynamics of human influenza A virus. *Nature* 453:615–19.

Smith, G. J. D., D. Vijaykrishna, J. Bahl, et al. 2009. Origins and evolutionary genomics of the 2009 swine-origin H1N1 influenza A epidemic. *Nature* 459:1122–25.

Taubenberger, J. K., and D. M. Morens. 2008. The pathology of influenza virus infections. *Annual Reviews of Pathology* 3:499–522.

RABBITS WITH HORNS

Bravo, I. G., and Á. Alonso. 2006. Phylogeny and evolution of papillomaviruses based on the E1 and E2 proteins. *Virus Genes* 34:249–62.

Doorbar, J. 2006. Molecular biology of human papillomavirus infection and cervical cancer. *Clinical Science* 110:525.

García-Vallvé, S., Á. Alonso, and I. G. Bravo. 2005. Papillomaviruses: Different genes have different histories. *Trends in Microbiology* 13:514–21.

García-Vallvé, S., J. R. Iglesias-Rozas, Á. Alonso, and I. G. Bravo. 2006. Different papillomaviruses have different repertoires of transcription factor binding sites: Convergence and divergence in the upstream regulatory region. *BMC Evolutionary Biology* 6:20.

Horvath, C. A. J., G. A. V. Boulet, V. M. Renoux, P. O. Delvenne, and J.-P. J. Bogers. 2010. Mechanisms of cell entry by human papillomaviruses: An overview. *Virology Journal* 7:11.

Martin, D., and J. S. Gutkind. 2008. Human tumor-associated viruses and new insights into the molecular mechanisms of cancer. *Oncogene* 27 (Suppl 2): S31–42.

Schiffman, M., R. Herrero, R. DeSalle, et al. 2005. The carcinogenicity of human papillomavirus types reflects viral evolution. *Virology* 337:76–84.

THE ENEMY OF OUR ENEMY

Brussow, H. 2005. Phage therapy: The *Escherichia coli* experience. *Microbiology* 151:2133.

Deresinski, S. 2009. Bacteriophage therapy: Exploiting smaller fleas. *Clinical Infectious Diseases* 48:1096–1101.

Sulakvelidze, A., Z. Alavidze, and J. G. Morris Jr. 2001. Bacteriophage therapy. *Antimicrobial Agents and Chemotherapy* 45:649.

Summers, W. C. 2001. Bacteriophage therapy. *Annual Reviews in Microbiology* 55:437–51.

THE INFECTED OCEAN

Angly, F. E., B. Felts, M. Breitbart, et al. 2006. The marine viromes of four oceanic regions. *PLoS Biology* 4 (11): e368.

Brussaard, C. P. D., S. W. Wilhelm, F. Thingstad, et al. 2008. Global-scale processes with a nanoscale drive: The role of marine viruses. *ISME Journal* 2:575–78.

Danovaro, R., A. Dell' Anno, C. Corinaldesi, et al. 2008. Major viral impact on the functioning of benthic deep-sea ecosystems. *Nature* 454:1084–87.

Desnues, C., B. Rodriguez-Brito, S. Rayhawk, et al. 2008. Biodiversity and biogeography of phages in modern stromatolites and thrombolites. *Nature* 452:340–43.

Rohwer, F., and R. Vega Thurber. 2009. Viruses manipulate the marine environment. *Nature* 459:207–12.

Suttle, C. A. 2007. Marine viruses—major players in the global ecosystem. *Nature Reviews Microbiology* 5:801–12.

Van Etten, J. L., L. C. Lane, and R. H. Meints. 1991. Viruses and viruslike particles of eukaryotic algae. *Microbiology and Molecular Biology Reviews* 55:586.

Williamson, S. J., S. C. Cary, K. E. Williamson, et al. 2008. Lysogenic virus-host interactions predominate at deep-sea diffuse-flow hydrothermal vents. *ISME Journal* 2:1112–21.

OUR INNER PARASITES

Blikstad, V., F. Benachenhou, G. O. Sperber, and J. Blomberg. 2008. Evolution of human endogenous retroviral sequences: A conceptual account. *Cellular and Molecular Life Sciences* 65:3348–65.

Dewannieux, M., F. Harper, A. Richaud, et al. 2006. Identification of an infectious progenitor for the multiple-copy HERV-K human endogenous retroelements. *Genome Research* 16:1548–56.

Jern, P., and J. M. Coffin. 2008. Effects of retroviruses on host genome function. *Annual Review of Genetics* 42:709–32.

Ruprecht, K., J. Mayer, M. Sauter, K. Roemer, and N. Mueller-Lantzsch. 2008. Endogenous retroviruses and cancer. *Cellular and Molecular Life Sciences* 65:3366–82.

Tarlinton, R., J. Meers, and P. Young. 2008. Biology and evolution of the endogenous koala retrovirus. *Cellular and Molecular Life Sciences* 65:3413–21.

Weiss, R. A. 2006. The discovery of endogenous retroviruses. *Retrovirology* 3:67.

THE YOUNG SCOURGE

Fan, H. 2011. *AIDS: Science and society*. 6th ed. Sudbury, MA: Jones and Bartlett.

Gilbert, M. T. P., A. Rambaut, G. Wlasiuk, T. J. Spira, A. E. Pitchenik, and M. Worobey. 2007. The emergence of HIV/AIDS in the Americas and beyond. *Proceedings of the National Academy of Sciences* 104:18566.

Keele, B. F. 2006. Chimpanzee reservoirs of pandemic and nonpandemic HIV-1. *Science* 313:523–26.

Montagnier, L. 2010. 25 Years after HIV discovery: Prospects for cure and vaccine. *Virology* 397:248–54.

Niewiadomska, A. M., and X.-F. Yu. 2009. Host restriction of HIV-1 by APOBEC3 and viral evasion through Vif. *Current Topics in Microbiology and Immunology* 339:1–25.

Worobey, M., M. Gemmel, D. E. Teuwen, et al. 2008. Direct evidence of extensive diversity of HIV-1 in Kinshasa by 1960. *Nature* 455:661–64.

BECOMING AN AMERICAN

Brault, A. C. 2009. Changing patterns of West Nile virus transmission: Altered vector competence and host susceptibility. *Veterinary Research* 40:43.

Diamond, M. S. 2009. Progress on the development of therapeutics against West Nile virus. *Antiviral Research* 83:214–27.

Gould, E. A., and S. Higgs. 2009. Impact of climate change and other factors on emerging arbovirus diseases. *Transactions of the Royal Society of Tropical Medicine and Hygiene* 103:109–21.

Hamer, G. L, U. D. Kitron, T. L. Goldberg, et al. 2009. Host selection by *Culex pipiens* mosquitoes and West Nile virus amplification. *American Journal of Tropical Medicine and Hygiene* 80:268.

Sfakianos, J. N. 2009. *West Nile virus.* 2nd ed. New York: Chelsea House.

Venkatesan M., and J. L. Rasgon. 2010. Population genetic data suggest a role for mosquito-mediated dispersal of West Nile virus across the western United States. *Molecular Ecology* 19:1573–84.

PREDICTING THE NEXT PLAGUE

Holmes, E. C., and A. Rambaut. 2004. Viral evolution and the emergence of SARS coronavirus. *Philosophical Transactions of the Royal Society B: Biological Sciences* 359:1059–65.

Parrish, C. R., E. C. Holmes, D. M. Morens, et al. 2008. Cross-species virus transmission and the emergence of new epidemic diseases. *Microbiology and Molecular Biology Reviews* 72:457–70.

Skowronski, D. M., C. Astell, R. C. Brunham, et al. 2005. Severe acute respiratory syndrome (SARS): A year in review. *Annual Review of Medicine* 56:357–81.

Wolfe, N. 2009. Preventing the next pandemic. *Scientific American*, April 2009, 76–81.

THE LONG GOODBYE

Hughes, A. L., S. Irausquin, and R. Friedman. 2010. The evolutionary biology of poxviruses. *Infection, Genetics and Evolution* 10:50–59.

Jacobs, B. L., J. O. Langland, K. V. Kibler, et al. 2009. Vaccinia virus vaccines: Past, present and future. *Antiviral Research* 84:1–13.

Kennedy, R. B., I. Ovsyannikova, and G. A. Poland. 2009. Smallpox vaccines for biodefense. *Vaccine* 27 (Suppl): D73–79.

Koplow, D. A. 2003. *Smallpox: The fight to eradicate a global scourge.* Berkeley: University of California Press.

McFadden, G. 2010. Killing a killer: What next for smallpox? *PLoS Pathogens* 6 (1): e1000727.

Shchelkunov, S. N. 2009. How long ago did smallpox virus emerge? *Archives of Virology* 154:1865–71.

EPILOGUE

Claverie, J.-M., and C. Abergel. 2009. Mimivirus and its virophage. *Annual Review of Genetics* 43:49–66.

Forterre, P. 2010. Defining life: The virus viewpoint. *Origins of Life and Evolution of Biospheres* 40:151–60.

Moreira, D., and C. Brochier-Armanet. 2008. Giant viruses, giant chimeras: The multiple evolutionary histories of mimivirus genes. *BMC Evolutionary Biology* 8:12.

Moreira, D., and P. Lopez-Garcia. 2009. Ten reasons to exclude viruses from the tree of life. *Nature Reviews Microbiology* 7:306–11.

Ogata, H., and J. M. Claverie. 2008. How to infect a mimivirus. *Science* 321:1305.

Raoult, D., and P. Forterre. 2008. Redefining viruses: Lessons from mimivirus. *Nature Reviews Microbiology* 6:315–19.

Raoult, D., B. La Scola, and R. Birtles. 2007. The discovery and characterization of mimivirus, the largest known virus and putative pneumonia agent. *Clinical Infectious Diseases* 45:95–102.

Credits

Introduction: tobacco mosaic viruses, © Dennis Kunkel Microscopy, Inc. Chapter 1: rhinovirus, copyright © 2010 Photo Researchers, Inc. (all rights reserved). Chapter 2: influenza virus, by Frederick Murphy, from the PHIL, courtesy of the CDC. Chapter 3: human papillomavirus, copyright © 2010 Photo Researchers, Inc. (all rights reserved). Chapter 4: bacteriophages, courtesy of Graham Colm. Chapter 5: marine phage, courtesy of Willie Wilson. Chapter 6: avian leukosis virus, courtesy of Dr. Venugopal Nair and Dr. Pippa Hawes, Bioimaging group, Institute for Animal Health. Chapter 7: human immunodeficiency virus, by P. Goldsmith, E. L. Feorino, E. L. Palmer, and W. R. McManus, from the PHIL, courtesy of the CDC. Chapter 8: West Nile virus, by P. E. Rollin, from the PHIL, courtesy of the CDC. Chapter 9: Ebola virus, by Cynthia Goldsmith, from the PHIL, courtesy of the CDC. Chapter 10: smallpox virus, by Frederick Murphy, from the PHIL, courtesy of the CDC. Epilogue: mimivirus, courtesy of Dr. Didier Raoult, Research Unit in Infectious and Tropical Emergent Diseases (URMITE).

Index

nonliving, 91–93; meaning of word, 3, 4, 94; origins of life and, 91; scope of, 33–34; size of, 4–5; varied locations of, 1–3. *See also specific viruses*

Weiss, Robin, 48–49
West Nile virus, 64–69
white blood cells: avian leukosis and, 54; HIV and, 54

Willner, Dana, 2–3
World Health Organization: SARS and, 76; smallpox and, 85, 87–88
World War I, bacteriophages and, 34, 36
World War II, bacteriophages and, 37
Worobey, Michael, 59, 60

zinc, common cold and, 11
Zur Hausen, Harald, 25